# NON-CONVENTIONAL-WEAPONS
# PROLIFERATION IN THE
# MIDDLE EAST

# NON-CONVENTIONAL-WEAPONS PROLIFERATION IN THE MIDDLE EAST

## Tackling the Spread of Nuclear, Chemical, and Biological Capabilities

*Edited by*

EFRAIM KARSH

MARTIN S. NAVIAS

*and*

PHILIP SABIN

CLARENDON PRESS · OXFORD
1993

Oxford University Press, Walton Street, Oxford OX2 6DP
Oxford New York Toronto
Delhi Bombay Calcutta Madras Karachi
Kuala Lumpur Singapore Hong Kong Tokyo
Nairobi Dar es Salaam Cape Town
Melbourne Auckland Madrid
and associated companies in
Berlin Ibadan

Oxford is a trade mark of Oxford University Press

Published in the United States
by Oxford University Press Inc., New York

British Library Cataloguing in Publication Data
Data available

Library of Congress Cataloging in Publication Data
Non-conventional-weapons proliferation in the Middle East: tackling
the spread of nuclear, chemical, and biological capabilities/
edited by Efraim Karsh, Martin Navias, and Philip Sabin.
Based on papers presented originally at a conference held by the
Dept. of War Studies at King's College, London, in Nov. 1991.
Includes index.
1. Middle East—Defenses—Congresses. 2. Arms race—Middle East—
Congresses. 3. Arms control—Middle East—Congresses. I. Karsh,
Efraim. II. Navias, Martin S. III. Sabin, Philip A. G.
uA832.N66 1993 355'.033056—dc20 93-9469
ISBN 0-19-827768-7

Typeset by Best-set Typesetter Ltd., Hong Kong
Printed in Great Britain
on acid-free paper by
Biddles Ltd., Guildford and King's Lynn

*In Memory of Avner Yaniv*

# ACKNOWLEDGEMENTS

The studies in this volume originated in a conference held by the Department of War Studies at King's College London in November 1991. They have since been revised to take account of the comments of other participants, and to encompass more recent developments in the field. The contributors include many of the leading international experts on proliferation and Middle Eastern security, from the United Kingdom, the United States, and the Middle East itself.

The editors would like to express their gratitude to those people and institutions who made this project possible. Foremost must come the Economic and Social Science Research Council and the UK Ministry of Defence, which funded the whole enterprise (though they take no responsibility for the views expressed in this volume). We are also grateful to all our contributors, for sticking to deadlines and hence ensuring prompt publication of the present volume. Thanks are due also to those who attended the conference and contributed many valuable thoughts, as well as to Marie-France Desjardins, Yossi Mekelberg, Andrew Rathmell, and Lee Willett for their help with the running of the conference. Finally, we are grateful to Oxford University Press for their kind co-operation throughout the publication process.

As this book went to press, the editors were sad to learn of the untimely death of Professor Avner Yaniv, a leading Israeli strategist and a contributor to this volume. His loss will be greatly felt by his friends and colleagues. It is to his memory that this volume is dedicated.

Efraim Karsh, Martin Navias, Philip Sabin

*London*
*July 1992*

# CONTENTS

# ABBREVIATIONS

| | |
|---|---|
| ACDA | Arms Control and Disarmament Agency |
| BTWC | Biological and Toxin Weapons Convention |
| CFE | Conventional Forces in Europe |
| CIA | Central Intelligence Agency |
| COCOM | Co-ordinating Committee for Multilateral Export Controls |
| CWC | Chemical Weapons Convention |
| DSAA | Department of Defense Security Assistance |
| EC | European Community |
| EMIS | Electromagnetic isotope separation |
| IAEA | International Atomic Energy Agency |
| IDF | Israel Defence Forces |
| IISS | International Institute for Strategic Studies |
| INF | Intermediate-range Nuclear Forces |
| MTCR | Missile Technology Control Regime |
| NATO | North Atlantic Treaty Organization |
| NBC | Nuclear, biological, and chemical |
| NPT | Non-Proliferation Treaty |
| OPEC | Organization of Petroleum Exporting Countries |
| PLO | Palestine Liberation Organization |
| SIPRI | Stockholm International Peace Research Institute |
| START | Strategic Arms Reduction Treaty |
| UN | United Nations |
| WEU | Western European Union |

# CONTRIBUTORS

LAWRENCE FREEDMAN is Professor and Head of the Department of War Studies, King's College, London. He has held research positions at Nuffield College, Oxford, and at the International Institute for Strategic Studies, and was Head of Policy Studies at the Royal Institute for International Affairs. He is the author of many books, including *Britain and Nuclear Weapons* (1981), *The Atlas of Global Strategy* (1985), *Britain and the Falklands War* (1988), *Signals of War* (1990), and (with Efraim Karsh) *The Gulf Conflict, 1990–1991: Diplomacy and War in the New World Order* (1993).

WILLIAM HOPKINSON is a civil servant in the UK Ministry of Defence. As head of the Ministry's Defence Arms Control Unit until 1990, he had a central role in the negotiation of the CFE agreement. His contribution to this volume was written during a sabbatical year as a Visiting Fellow in Cambridge University's Global Security Programs, during which he also produced a monograph entitled *Changing Options: British Defence and Global Security* (1992).

EFRAIM KARSH is Reader at the Department of War Studies, King's College, London. He has held teaching and/or research posts at Columbia University, the International Institute for Strategic Studies (London), the Kennan Institute for Advanced Russian Studies (Washington, DC), and the Jaffee Center for Strategic Studies (Tel Aviv). His many publications include *The Iran–Iraq War: A Military Analysis* (1987), *Neutrality and Small States* (1988), *The Soviet Union and Syria* (1988), *Soviet Policy towards Syria since 1970* (1991), *Saddam Hussein: A Political Biography* (1991), (with Lawrence Freedman) *The Gulf Conflict, 1990–1991: Diplomacy and War in the New World Order* (1993).

GEOFFREY KEMP is a Senior Associate at the Carnegie Endowment for International Peace. He served in the White House on the National Security Council Staff from 1981 to 1984 and was Senior Director for Near East and South Asian Affairs. His most recent publications include *The Control of the Middle East Arms Race* (1991).

MARTIN NAVIAS is a lecturer in the Department of War Studies, King's College, London, where he specializes in proliferation issues and the international arms trade. His previous publications include *Nuclear Weapons and British Strategic Planning, 1955–1958* (1991) and *Ballistic Missile Proliferation in the Third World* (Adelphi Papers, 252; summer

1990). He has recently completed *Going Ballistic: The Build-up of Missiles in the Middle East* (1993).

STEPHANIE NEUMAN is a Senior Research Scholar and Director of the Comparative Defense Studies Program at Columbia University, where she also teaches on Third World security issues. Her most recent publications include *Assistance in Recent Wars* (1986) and *Wars in the Third World: A Synthesis of Lessons Learned* (1993).

GRAHAM PEARSON CB is Director General of the Chemical and Biological Defence Establishment at Porton Down, UK.

JULIAN PERRY ROBINSON, a chemist and lawyer by training, is a senior fellow of the Science Policy Research Unit, University of Sussex, England, where he heads the Military Technology and Arms Limitation research group. He had previously held research appointments at the Stockholm International Peace Research Institute (SIPRI), the Free University of Berlin, and the Center of International Affairs at Harvard University. At SIPRI during 1968–71 he wrote much of the six-volume study *The Problem of Chemical and Biological Warfare*, and during 1982–6 was the founding editor of the series *SIPRI Chemical and Biological Warfare Studies*. With Matthew Meselson of Harvard University, he directs the Harvard–Sussex Program on CBW Armament and Arms Limitation, co-editing the program's journal, *Chemical Weapons Convention Bulletin*, published quarterly in Washington.

PHILIP SABIN has held research fellowships at Harvard University and the International Institute for Strategic Studies, and has been a lecturer in the Department of War Studies at King's College, London, since 1986. His publications, include a book on *The Third World War Scare in Britain* (1986), Adelphi Paper on *Perceptions and Symbolism in Nuclear Force Planning* (1987), and *British Strategic Priorities in the 1990s* (1990), and articles on a wide variety of defence topics including 'Escalation in the Iran–Iraq War' (1989).

YEZID SAYIGH is MacArthur Scholar and Research Fellow at St Antony's College, Oxford, working on the international relations of the developing countries and Third World security. He is also an adviser to the Palestinian delegation in the bilateral peace talks and co-ordinator of the Palestinian team for arms control and regional security within the multilateral talks. His most recent publications include *Confronting the 1990s: Security in the Developing Countries* (1990), *Arab Military Industry: Capability, Performance and Impact* (1992), and *The Palestinian Armed Struggle since 1949* (forthcoming in 1993).

LEONARD SPECTOR is the Director of the Carnegie Endowment's Nuclear Non-Proliferation Project. Previously he worked for the Nuclear Regulatory Commission and then as Chief Counsel of the Senate Energy and Nuclear Proliferation Subcommittee. He has been the author of the Carnegie Endowment's series on nuclear proliferation. The most recent edition, *Nuclear Ambitions* (written with the assistance of Jacqueline Smith), was published by Westview Press in 1990.

AVNER YANIV was, until his untimely death in 1992, a Professor of Political Science and Vice President of the University of Haifa. A specialist on Israeli defence and foreign policy, his many publications include *Dilemmas of Security: Politics, Strategy, and the Israeli experience in Lebanon* (1987), *Deterrence Without the Bomb* (1987).

# Introduction

## EFRAIM KARSH, MARTIN NAVIAS, AND PHILIP SABIN

THE 1980s witnessed the escalation of the arms race in the Middle East on to a higher qualitative level as a result of the widespread proliferation of non-conventional instruments of warfare and technologies. True, the spectre of non-conventional-weapons proliferation is not completely novel to the Middle East. The Egyptians had already employed gas in the Yemen civil war in the 1960s, while Israeli nuclear-development efforts are alleged to have begun a decade earlier. However, the far greater scope of the recent proliferation, its intensity and breadth and its higher level of technological sophistication, has rendered the issue of much greater political and military significance.

The Iran–Iraq War not only underlined the horrific nature of non-conventional warfare but also served to undermine a number of crucial thresholds relating to the uses of non-conventional weaponry, thereby making future employment of such systems seem increasingly probable. Concomitantly, the revelations by Israeli nuclear technician Mordechai Vanunu concerning Israel's burgeoning nuclear capabilities, Iraq's grim determination to acquire the bomb, and the disintegration of the Soviet Union have made the nuclearization of the Middle East closer than ever. Since proliferation in this part of the world is believed to pose the greatest challenge to international security and stability, it is hardly surprising that the Middle East has been subjected to emerging efforts aimed at constraining and controlling the spread of non-conventional ordnance.

While estimates vary, it is generally accepted that Israel possesses a sizeable and sophisticated nuclear arsenal. In addition, Iran continues to pour substantial sums into building up its own nuclear force, while Algeria has purchased a nuclear facility capable of manufacturing weapons-grade uranium. Until the 1991 Gulf War Iraq was believed to be a few years away from producing a nuclear weapon, but revelations after the war (mainly by defecting Iraqi nuclear scientists) indicated that the Iraqi nuclear programme had been far more advanced and extensive than previously assumed.

Whether the war and the consequent international effort to dismantle Iraq's non-conventional programme will be fully successful still remains to be seen: over a year after the termination of hostilities, Iraqi and UN officials were still haggling over the size of Iraq's nuclear programme, its whereabouts, and the commitment of Baghdad to its destruction. What is clear, however, is that the difficulties of ascertaining the parameters of Iraq's nuclear efforts underline the problems associated with determining the nature and extent of not only nuclear proliferation in Iraq but the spread of other forms of non-conventional weaponry in the region.

Chemical weapons are believed to be in the arsenals of a number of Middle Eastern countries. Iraq was found to possess an enormous quantity of chemical ordnance. Israel, no doubt, has a sophisticated chemical arsenal, though there is admittedly little information on this subject. Since the Iran–Iraq War, Tehran has sought to upgrade its chemical-weapons capabilities, while Syria is known to have developed an advanced chemical-production infrastructure and an extensive range of chemical munitions. Egypt, too, is thought to possess chemical weapons, and, at the desert town of Rabta, Libya may have the largest chemical-weapons production facility in the world.

Whereas chemical weapons have failed to live up to the description of being the 'poor man's atom bomb', biological weapons present a different problem altogether. The many uncertainties surrounding biological warfare, associated with the lack of experience with regard to its employment and the wide variety of biological munitions potentially available, render predictions in this field extremely difficult to make. Whatever the case, Middle Eastern countries have shown great interest in at least investigating the military utility of such substances. Despite many denials, Iraq appears to have been undertaking a biological-weapons programme. Iran and Syria, too, are thought to possess or to be involved in the development of these munitions. Egypt and Israel, for their part, are sometimes mentioned as either possessing biological weapons or at least having the capability to develop them.

The acquisition by Middle Eastern states of non-conventional weapons has been accompanied by a search for top-grade conventional weaponry, including items that could be employed to deliver non-conventional munitions over a wide circumference. Thus the Middle East is home to some of the world's most sophisticated fighter aircraft and also ballistic missiles. The latter have been of great concern, because of the great difficulties encountered in destroying them once launched—as illustrated by the mixed performance of the Patriot anti-missile missile during the 1991 Gulf War. Not surprisingly, the war has accelerated regional attempts to develop and deploy a wide range of surface-to-surface ballistic missiles. Despite international efforts—or at least declaratory expressions—to control the flow of arms to the Middle East, it resumed with consistent intensity in the wake of the Gulf War.

A major concern among many analysts over the past few years has been that the spread of nuclear, chemical, and biological weapons has not been accompanied by either practical steps or conceptual theorizing aimed at preventing, or at least limiting, the employment of these highly destructive weapons. This concern has been further compounded by the view that, due to the relative coarseness and insensitivity to casualties and suffering shown by a number of Middle Eastern countries, they may be more prone than others to the future employment of weapons of mass destruction. The speed at which proliferation has taken place has also meant that concepts of deterrence and strategies of restraint have not been firmly embedded in national decision-making processes. Finally, the extreme objectives of various regional protagonists open up the possibility that scenarios could well arise when weapons of mass destruction are resorted to in desperate situations.

Certainly, the implications of this proliferation are unlikely to remain restricted to regional conflicts alone. The 1991 Gulf War demonstrated the apprehension caused in the West by the possibility of having to fight against an enemy armed with chemical weapons. Had Saddam Hussein deployed nuclear ordnance, then it is quite possible that allied policy options would have required a serious reconsideration. Future power projections into this region could well be complicated by a foe threatening to employ non-conventional weapons. Last but not least, in the next few years it is not inconceivable that Middle Eastern proliferators could threaten states outside the region with nuclear, chemical, or biological weapons. This is certainly part of the reason why extra-regional powers are watching so closely the correlation between non-conventional and ballistic-missile proliferation in the Middle East.

Middle Eastern proliferation has also placed severe strains on the credibility of international agreements aimed at limiting the spread of the weapons in question. Iraq, for example, has long been a member of the NPT, yet it has consistently sought to develop nuclear weapons. The failure of this control regime to identify and ultimately to halt the Iraqi programme has underlined its severe limitations. With Iraq's programme now stymied, it is Iran's actions—Tehran too is a member of the NPT—which serve to highlight the inadequacies of the arms-control regime. Concomitantly, Israel's steadfast refusal to join the NPT also underlines the inability of this international framework fully to curb proliferation in the region.

The difficulties revealed with regard to the efficacy of international control regimes aimed at preventing nuclear proliferation have been mirrored by the failure of the 1925 Geneva Protocols to halt the manufacture of chemical weapons (which it did not actually seek to do) and their employment (which it did). The draft Chemical Weapons Convention adopted by the Conference on Disarmament in September 1992 is more ambitious but faces a daunting task in halting the build-up of chemical stockpiles in the Middle East. Clearly the issue of enhancing, expanding, and enforcing non-conventional control regimes will remain a crucial policy issue throughout the 1990s.

Following the successful ejection of Iraq from Kuwait by a US-led international coalition, President Bush unveiled his Middle Eastern arms-control policy, which included a heavy focus on weapons of mass destruction. Announced on 29 May 1992, the President's scheme called for an international ban on chemical and biological weapons; a verifiable ban on the acquisition and production of plutonium and enriched uranium which could be employed in the production of chemical weapons; an attempt to constrain the supply of the most dangerous conventional weapons, and a freeze leading to an eventual ban on the testing, manufacture, and purchase of surface-to-surface ballistic missiles. These proposals soon received firm backing from the five permanent members of the UN Security Council and a wide variety of other international forums and bodies.

Yet, despite these best of intentions, it is clear that there is only limited regional (and, to a lesser extent, international) enthusiasm for both conventional and non-conventional arms control. Declaratory policy aside, the desire of Middle Eastern states for weapons of mass destruction remains strong, while the producers of such weapons remain as keen as ever to export their products. The problem, therefore, is likely to remain with us for some time to come, and its essence and consequences are discussed in the present volume.

By way of addressing these questions, the book has been divided into four parts. The first part discusses the historical experience of missile and non-conventional-weapons employment, especially in the Middle East. Philip Sabin opens with a broad historical perspective on the use or non-use of non-conventional weapons, and discusses why states possessing such weapons have very often held back from employing them in war. Efraim Karsh follows with a detailed examination of the Iran–Iraq War, arguing that the use of missiles and chemical weapons in that conflict constituted 'rational ruthlessness' and was subject to definite limitations and constraints. Martin Navias then examines the 1991 Gulf War, arguing that, even though missiles and non-conventional weapons played a very limited military role, their *political* importance was much greater, and hence they remain critical to the Middle Eastern security equation.

Part Two assesses the current state of non-conventional-weapons proliferation in the region, and the prospects for the future. Examining the spread of chemical weapons in the Middle East, Julian Perry Robinson casts doubt on their military utility and on the inevitability of continued proliferation. A much gloomier view of the potential utility of germ warfare is taken by Graham S. Pearson, who discusses biological weapons, particularly the recent findings in this respect in Iraq. Finally, Leonard Spector reviews the state of nuclear proliferation in the Middle East, and assesses the viability of the existing non-proliferation regime in the face of intensifying nuclear aspirations on the part of several states in the region.

After this discussion of the actual state of non-conventional proliferation in the Middle East, Part Three addresses its strategic impact in the context of the 'new world order' and the end of the Cold War. Lawrence Freedman starts with an examination of how Middle Eastern proliferation affects and is likely to be affected by the new world order proclaimed by President Bush. He is followed by Yezid Sayigh, who analyses the impact of proliferation on Middle Eastern stability, arguing that, unless proliferation is brought under control, it will escalate tensions and provoke further armed conflict. Avner Yaniv then focuses on the specific issue of the Arab–Israeli confrontation, exploring the applicability of Cold War concepts of deterrence in this very different situation.

Finally, Part Four examines the prospects for restraining Middle Eastern proliferation through measures of arms control. It begins with an examination by Bill Hopkinson of whether Middle Eastern proliferation could be tackled by negotiated arms control or by supplier restraints, and whether any lessons from recent experience in Europe are applicable in this case. Geoffrey Kemp then explores the arms-control option further, and addresses the all-important linkage with the Arab–Israeli peace process. The book is concluded with a chapter by Stephanie Neuman, assessing the prospects for restraints on arms transfers from outside powers to the Middle East, and drawing attention to the blurring of the distinction between conventional and non-conventional-weapons capabilities.

# PART ONE

# HISTORICAL EXPERIENCE

# 1

# Restraints on Chemical, Biological, and Nuclear Use: Some Lessons from History

## PHILIP SABIN

DURING the 1991 Gulf War, the spectre of an Iraqi chemical attack caused considerable anxiety among Iraq's opponents. Coalition forces undertook wide-ranging defensive precautions against this threat, despite the inconvenience of using equipment designed for European warfare in the hotter climate of the Persian Gulf. Israel urged its citizens to shelter in upper rooms rather than basements to mitigate the impact of possible chemical warheads on Iraqi Scud missiles, even though this meant sacrificing some protection against conventional attack.[1] In the event, however, Iraq did not use chemical weapons during the Gulf War, despite possessing at least some capability to do so.[2]

One of the reasons why the US-led coalition decided to tackle Iraq militarily rather than relying on longer-term solutions such as sanctions was a fear that Iraq might soon develop a nuclear and biological capability alongside its existing stockpile of chemical weapons. UN inspections after the war have proved that Iraq was indeed researching biological weapons and was well on the way to acquiring a rudimentary atomic arsenal.[3] However, the fact that Iraq did not use its chemical weapons during the 1991 Gulf War raises the question of whether it would have been any readier

---

[1] Since chemical warheads could have caused many more casualties in the right conditions, the Israeli decision was probably a rational one as long as chemical attack could not be ruled out. See S. Fetter, 'Ballistic Missiles and Weapons of Mass Destruction: What is the Threat? What Should be Done?', *International Security* (summer 1991), 5–42.

[2] See Martin Navias, Ch. 3, this volume.

[3] See Graham S. Pearson and Leonard Spector, Chs. 5 and 6, this volume.

to employ atomic or biological weapons in any future armed confrontation.

Because the 1991 Gulf War ended as it did, questions such as these have tended to take second place to discussions of how to keep such weapons out of the hands of states like Iraq in the first place. However, the destruction of Iraq's NBC capabilities under UN supervision does not solve the problem that other Middle Eastern states such as Israel and Syria already possess very significant capabilities of this kind. The 1992 CWC should help to alleviate at least part of this problem, but to expect arms control to produce a general and lasting abandonment of NBC weapons throughout the Middle East seems unduly optimistic.

This being the case, it is worth paying at least some attention to the question of why Middle Eastern states in the future might hold back from using NBC weapons in warfare, even if they possess the requisite capabilities. Without a proper understanding of the dynamics of restraint, policy-makers will be less able to head off NBC use in future Middle Eastern conflicts or to defuse the incentives which might produce NBC proliferation. The result could be undue reliance on coercion to prevent certain regional powers from acquiring NBC capabilities, even if this requires preventive military strikes similar to Israel's raid on the Iraqi nuclear plant in 1981.[4]

NBC weapons may, of course, have considerable strategic impact even without being directly employed in war. The mere *threat* of their use may impose caution on other powers, by deterring them from initiating or expanding hostilities or by forcing them to take debilitating practical precautions, as happened during the 1991 Gulf War. However, the credibility of this threat rests ultimately on the willingness of the state in question to *use* its NBC weapons should appropriate circumstances arise, and it is this willingness which will be examined here, using the evidence provided by historical experience.

## THE RELEVANCE OF HISTORICAL PRECEDENT

There have been numerous conflicts over the past eighty years in which NBC weapons have been used, or have been held back,

---

[4] See A. Perlmutter, M. Handel, and U. Bar-Joseph, *Two Minutes over Baghdad* (London: Corgi, 1982).

despite being available to at least one belligerent. The danger is that any lessons which might be drawn from this experience as regards future NBC warfare in the Middle East might ignore significant contextual differences which make the comparison invalid. However, the range of historical precedent is sufficiently broad for this problem of generalization not to prove insuperable.

Although our most detailed knowledge of decision-making about NBC employment stems from the UK and US democracies, we also know a significant amount about the thinking of authoritarian states such as Imperial and Nazi Germany. Japanese policy illuminates a different cultural perspective, as do the policies of Middle Eastern states themselves. Although judgements about the past policies of states like Egypt and Iraq as regards chemical-weapons employment must unfortunately remain largely circumstantial, there is enough information to judge whether what we already know of NBC warfare in the Middle East is in accordance with patterns observed elsewhere.

This chapter will address six specific aspects of the issues concerned. First, it will assess the range of strategic circumstances in which states consider employing NBC weapons. Secondly, it will discuss how far the use of such weapons has been restrained by fear that the enemy would respond in kind. Thirdly, it will examine the impact of thresholds and limitations within NBC warfare. Fourthly, it will assess the role of cost–benefit analysis at the military level. Fifthly, it will address the issue of cultural aversion to the particular characteristics of NBC weapons. Finally, it will discuss the impact of political considerations such as international agreements and world opinion. The paper will conclude by assessing how likely it is that NBC capabilities in the Middle East will remain unused, even if disarmament and non-proliferation fail.

## CIRCUMSTANCES OF USE

The historical record does suggest one common feature regarding the strategic circumstances in which NBC weapons have been employed—namely, that such weapons have been introduced some time after the outbreak of wars rather than being employed from the outset. However, it is not possible to generalize further—for example, by concluding that NBC weapons have only been used in total war or as a means of staving off impending defeat. Instead, the circumstances in which such weapons have been employed vary widely, suggesting that the principal determinants of use or restraint lie elsewhere.

States do sometimes resort to NBC weaponry when their backs are to the wall, as did Iraq when facing Iranian human-wave assaults during the Iran–Iraq War.[5] The United Kingdom might well have used gas had the Germans launched a seaborne invasion in 1940, and Israel is reported to have readied nuclear warheads during the dark days of the 1973 Arab–Israeli War.[6] However, states do not necessarily resort to NBC weapons in this way, even if they are on the brink of enemy occupation. France fell in 1940 without using its chemical arsenal, as did Germany in its turn in 1945, whether or not Hitler gave orders at the last for a chemical Götterdämmerung.[7]

Historically, states have been just as likely to initiate NBC warfare for *offensive* purposes, either to break a stalemate or to administer a *coup de grâce* to a defeated opponent. The first use of lethal chemicals in the First World War was intended to overcome the stagnation of the trench lines and allow the elimination of the Ypres salient, though there were few hopes that a strategically decisive result could be achieved.[8] The use of atomic bombs against Japan at the end of the Second World War had an even clearer offensive purpose, being designed to precipitate a Japanese surrender through what the then Secretary of War Henry Stimson later described as 'a tremendous shock which would carry convincing proof of our power to destroy the Empire'.[9]

Nor has NBC employment always taken place in the context of total war between major powers. Chemical weapons have been used at least as often by expeditionary forces in more peripheral conflicts, as by the British in Russia in 1919, the Italians in Ethiopia and the Japanese in China in the 1930s, the Americans in

---

[5] See Efraim Karsh, Ch. 2, this volume.
[6] See V. Utgoff, *The Challenge of Chemical Weapons* (London: Macmillan, 1990), 37–42; H. Catudal, *Israel's Nuclear Weaponry* (London: Grey Seal, 1990), 54.
[7] See Utgoff, *Challenge of Chemical Weapons*, 34–7, 55–6.
[8] See U. Trumpener, 'The Road to Ypres: The Beginnings of Gas Warfare in World War I', *Journal of Modern History* (Sept. 1975), 460–80, esp. pp. 474–6.
[9] See L. Freedman, 'The Strategy of Hiroshima', *Journal of Strategic Studies* (May 1978), 87.

Vietnam and the Egyptians in Yemen in the 1960s.[10] Although nuclear weapons have not been used again since 1945, the United States is known to have considered their employment in peripheral conflicts in Asia as well as in the context of a Third World War between East and West.[11]

The fact that NBC weapons have not so far been used at the outset of a war does not mean that this will not occur in some future conflict in the Middle East or elsewhere, but it does suggest that a sudden nuclear or chemical strike may be less of a risk than the escalation of hostilities which begin at the conventional level. Since historical precedent gives little guidance about whether such escalation is more likely to be initiated by the attacker or the defender, or in a central or peripheral conflict, attention will be turned now to other determinants of use or restraint, as revealed by historical experience.

### FEAR OF ENEMY RETALIATION

The pattern of NBC warfare over the past eighty years suggests strongly that the most important disincentive to use these weapons has been fear that the enemy would respond in kind. Most instances of NBC use have been against opponents incapable of effective retaliation, whereas in other conflicts—such as the Second World War and the 1991 Gulf War—chemical weapons have been held back, despite being available to both sides. However, it would be wrong to deduce too much from this simple correlation, without a closer look at the processes involved.

For mutual deterrence alone to restrain the employment of NBC weapons, *both* sides must see it as less in their interest to use such weapons than to suffer their use in return. Since asymmetries in capability and vulnerability mean that NBC warfare is likely to be more damaging to one side than the other, the side which faces less damage is unlikely to be deterred by relative damage assessments alone. Unless other factors also come into play, the tendency will be for that side to initiate use of NBC weapons to gain a military advantage, as happens in other dimensions of warfare.

---

[10] See SIPRI, *The Problem of Chemical and Biological Warfare*, i. *The Rise of CB Weapons* (Stockholm: Almqvist & Wiksell, 1971), 141–52, 159–210.

[11] See R. Betts, *Nuclear Blackmail and Nuclear Balance* (Washington DC: Brookings, 1987), 31–62, 68–79.

This is what took place in the First World War, producing the one major instance of bilateral chemical conflict. In early 1915 Falkenhayn, the Chief of the German General Staff, decided to initiate the use of chlorine gas on the Western Front. One of his commanders objected that the enemy would adopt the device and, given the prevailing winds, could release gas against the Germans ten times as often. However, Falkenhayn and his scientific adviser Fritz Haber replied that the Allied chemical industry was 'simply not capable of producing gas in the quantity needed'.[12] Here is a clear case of the risk of retaliation failing to deter, because one side believed that bilateral chemical warfare would be to its own advantage.

Three factors prevented a similar breakdown of chemical deterrence during the Second World War. First, poor intelligence led to widespread overestimation of enemy readiness for chemical warfare. Germany in fact enjoyed a monopoly in the field of nerve gas, but suffered from a chemical 'inferiority complex' because it assumed that its opponents had been forging ahead with developments while it had been restricted by the Treaty of Versailles. The Allies, although ignorant of nerve gas, overestimated Axis chemical preparedness in other areas, and were all too conscious of the deficiencies in their own preparations. Hence, no belligerent felt confident enough in its own superiority to initiate a chemical campaign.[13]

The second important factor was that, as states gained the upper hand militarily, their advancing forces became more vulnerable to disruption by chemical attack. The commander of German chemical troops, General Ochsner, explained that gas was not employed in the Nazi *blitzkrieg* operations because 'The use of chemical agents could only have reduced the speed in operations of this nature.'[14] Later in the war, a major disincentive to Allied use of chemical weapons was the vulnerability of amphibious operations such as the Normandy landings or the planned invasion of Japan to chemical retaliation. This led to a dynamic balance in which the losing side was deterred from using chemical weapons

[12] See Trumpener, 'The Road to Ypres', 470–3.
[13] See SIPRI, *Problem of Chemical and Biological Warfare*, i. 302–16.
[14] H. Ochsner, *History of German Chemical Warfare in World War II*, pt. I. *The Military Aspect* (Washington DC: US Chemical Corps Historical Study no. 2, 1949), 21.

by enemy air superiority, while the winning side was deterred by the vulnerability of its conventional military operations.[15]

The third factor which reinforced chemical deterrence during the Second World War was the fact that the damage which it would entail came to seem unacceptable to the belligerents, even if the enemy might be made to suffer more. In part this was the result of actual experience of the horrors of gas warfare in the First World War, but it stemmed more from the fact that developments in air power extended the area of risk from front-line troops to civilian populations at home. The spectre of gas bombardment terrified civilians in the inter-war period, and the vulnerability of their cities prompted all belligerents except the United States to tread very carefully so as to avoid provoking a chemical war.[16]

This last factor has become the major stabilizing influence on *nuclear* deterrence over the past four decades. Even in the 1950s and early 1960s, when US nuclear superiority over the Soviet Union was overwhelming and when a US first strike might have destroyed a large proportion of the Soviet Union's nuclear forces on the ground, US decision-makers were deterred by thoughts of the *absolute* damage that a handful of Soviet weapons might cause if they reached the United States.[17] Deterrence became all the stronger when the Soviet nuclear build-up produced a situation of 'Mutually Assured Destruction'.

We shall never know if either side would have been prepared to cross the nuclear threshold had the East–West confrontation actually erupted into open warfare. Both blocs proclaimed their willingness to use nuclear weapons if required, but many former decision-makers have expressed their doubts about whether this would actually have occurred.[18] The use of *chemical* weapons seems rather more plausible, given the Warsaw Pact's extensive preparations in this field, though even here it is far from clear that mutual deterrence would have failed.[19]

[15] See Utgoff, *Challenge of Chemical Weapons*, Ch. 3.

[16] See SIPRI, *Problem of Chemical and Biological Warfare*, i. 314–15.

[17] See R. Betts, 'A Nuclear Golden Age? The Balance before Parity', *International Security* (winter 1986–7).

[18] See, e.g., H. Kissinger, 'NATO: The Next Thirty Years', *Survival* (Nov.–Dec. 1979), 264–8; M. Bundy, G. Kennan, R. McNamara, and G. Smith, 'Nuclear Weapons and the Atlantic Alliance', *Foreign Affairs*, (spring 1982), 753–68.

[19] See E. Spiers, *Chemical Warfare* (London: Macmillan, 1986), chs. 6, 7.

What does all this suggest about the stability of NBC deterrence in the Middle East? One hopeful conclusion is that the growing mutual vulnerability of civilian populations in the region may make employment of NBC weapons due to a perceived military superiority in such weapons less likely. The precedents of Nazi Germany, the Soviet Union, and Ba'athist Iraq suggest that authoritarian states can be just as sensitive to these civilian vulnerabilities as democracies. However, the stakes should deterrence fail are becoming correspondingly high, and states may well decide to launch conventional attacks upon their opponent's NBC capabilities in the event of conflict, rather than rely for their survival entirely on the enemy's own restraint.

### THRESHOLDS AND LIMITATIONS

Crucial to the possibility of restraint in NBC warfare has been the fact that NBC weapons are sufficiently distinct from other arms for there to be a fairly clear threshold defining their use or non-use.[20] However, there are also certain other thresholds *within* NBC warfare. Historically, these have had a double-edged effect. On the one hand, they have limited the consequences when NBC warfare has broken out; but, on the other hand, they have undermined the disincentive to use NBC weapons in the first place.

An important reason for the non-use of chemical weapons in the Second World War was that few states believed that internal thresholds would prove sustainable. In Julian Perry Robinson's words, 'each belligerent came to realize that retaliation could well be escalatory: a chemical mortar action in some distant combat theatre, even with irritant-agent projectiles, might be met by the gas-bombing of a capital city'.[21]

The exception to this pattern was Japan, which continued to use poison gas in China, despite President Roosevelt's hollow warning in June 1942 that, 'if Japan persists in this inhumane form of warfare against China . . . such actions will be regarded by this Government as though taken against the United States, and retaliation in kind and in full measure will be meted out'.[22] Only when

---

[20] On the importance of such thresholds in controlling escalation, see T. Schelling, *Arms and Influence* (New Haven, Conn.: Yale University Press, 1966), 131–41.

[21] SIPRI, *Problem of Chemical and Biological Warfare*, i. 335.

[22] See Utgoff, *Challenge of Chemical Weapons*, 29–32.

the United States became better able to carry out this threat did Japan desist.

Despite Japan's success in avoiding US retaliation, both Eastern and Western states continued to tread carefully during the ensuing Cold War, lest their use of NBC weapons against a regional antagonist lead to direct intervention or the supply of retaliatory capabilities by an opposing great power. When UN use of atomic weapons in Korea was mooted in March 1953, US Army Chief Lawton Collins suggested that, 'Before we use them we had better look to our air defense. Right now we present ideal targets for atomic weapons in Pusan and Inchon.'[23] However, when chemical weapons were actually used by the United States in Vietnam and, allegedly, by the Soviet Union in Afghanistan, this did not prompt the opposing superpower to provide comparable retaliatory capabilities (not least because these would have been of little use to guerrilla forces).[24]

There have been several instances of chemical weapons being used against front-line troops, but not against the enemy civilian population. In the First World War this was due partly to the lack of suitable weapons, but in the Vietnam and Iran–Iraq wars it was more a product of deliberate restraint.[25] Another common phenomenon was for chemical employment to begin with irritant rather than lethal agents. Both sides used irritants in the early months of the First World War, and Iraq employed tear gas against Iranian forces in July 1982 before progressing to the use of mustard gas some months later.[26] However, only in Vietnam did chemical warfare remain confined to large-scale use of irritants and defoliants, without escalating in due course to the employment of lethal gases.

The thresholds between the three different types of NBC weapons have proved reasonably sustainable over the past fifty years, and it is interesting that their use has not always occurred according to a strict hierarchy of destructiveness. The United States used atomic bombs against Japan without initiating chemical or biological warfare, and later employed chemical weapons in Vietnam without any serious thought of resorting to nuclear or

[23] See Betts, *Nuclear Blackmail and Nuclear Balance*, 39.

[24] See E. Spiers, *Chemical Weaponry: A Continuing Challenge* (London: Macmillan, 1989), ch. 5.

[25] Ibid.; SIPRI, *Problem of Chemical and Biological Warfare*, i. 98–9.

[26] See Trumpener, 'The Road to Ypres', and Efraim Karsh, Ch. 2, this volume.

biological techniques. However, it should be noted that few other states which have used chemical weapons have had access to other NBC capabilities, and that Japan does seem to have experimented with biological- as well as chemical-weapons employment during the Second World War.[27]

These reservations apart, the historical record suggests that thresholds within NBC warfare may possess a more significant potential for restraint than many in the Second World War seem to have believed. If NBC weapons are used in the Middle East, the chances are that the belligerents will observe certain limitations, rather than employing all their capabilities without restraint. Attention will be turned now to what other factors besides fear of retaliation might cause Middle Eastern states to limit or forgo NBC operations in this way.

## MILITARY COST-EFFECTIVENESS

A major determinant of whether and how NBC weapons have been used in the past seems to have been the perceived military utility of such weapons compared to conventional munitions. Also important has been the mundane issue of logistics, with the lack of sufficient NBC stocks and defensive equipment immediately to hand often helping to dissuade states from initiating NBC warfare in response to shifting military opportunities and challenges. These two sets of issues will now be examined in turn.

Where chemical-weapons use has been initiated on the battlefield, this has often been to solve a particular tactical problem. German experiments with chemical weapons in 1914 were prompted by the shortage of high-explosive shells, and by complaints from the Front that these conventional shells were proving ineffective against enemy trenches.[28] US use of irritant agents in Vietnam was intended initially to help deal with Viet Cong forces intermingled with civilian populations and later became a means of flushing out cave defences so that the enemy could be killed by conventional firepower.[29]

---

[27] On this and other alleged instances of biological warfare, see SIPRI, *Problem of Chemical and Biological Warfare*, i. 111–24, 214–30, 332–3, 342–7.

[28] See Trumpener, 'The Road to Ypres', 464.

[29] See V. Adams, *Chemical Warfare: Chemical Disarmament* (London: Macmillan, 1989), 80–4.

Another major incentive for the use of gas has been the vulnerability of opposing forces to its physical and psychological effects. This helps to explain Italian chemical attacks in Ethiopia, and Egyptian chemical attacks in the Yemen, in both cases against lightly clad tribesmen lacking any form of protection.[30] Alleged Vietnamese chemical attacks in Laos and Kampuchea, and alleged Soviet use of chemicals in Afghanistan, would fit the same pattern. Iraq found gas a useful counter to human-wave attacks by Iranian infantry, and later used it against even less-well-protected Kurdish tribesmen in the north of the country.[31]

Where such specific problems or vulnerabilities did not exist, chemical weapons have often been seen as too difficult to handle and too unpredictable in their effects for their use to be worthwhile. As Julian Perry Robinson points out elsewhere in this volume, it is easier for a technically advanced power to protect its troops from chemical weapons (albeit at some cost in effectiveness) than from modern conventional munitions.[32] Civilian populations are harder to protect, but in the Second World War the delivery of chemicals by aircraft or missile was seen as less cost-effective against protected populations than the use of an equivalent weight of high explosive.[33] Biological weapons could cause much greater civilian casualties in the right circumstances, but they have tended to be viewed in the past as even less predictable and tractable instruments for military purposes than chemical arms.[34]

Nuclear weapons differ in that they clearly merit the appellation 'weapons of mass destruction', being several orders of magnitude more destructive than conventional munitions against large area targets such as cities, and being much more difficult than chemical weapons to defend against. However, the unwieldiness of nuclear weapons against targets other than cities has been a real constraint on their military employment.

General Lawton Collins noted in March 1953 that he was 'very sceptical about the value of using atomic weapons tactically in Korea', and President Eisenhower expressed similar scepticism about the utility of nuclear weapons to relieve the siege of Dien

---

[30] See Spiers, *Chemical Weaponry*, chs. 3, 5.
[31] Ibid., ch. 5.
[32] See Julian Perry Robinson, Ch. 4, this volume.
[33] SIPRI, *Problems of Chemical and Biological Warfare*, i. 100–2.
[34] See Graham Pearson, Ch. 5, this volume.

Bien Phu the following year.[35] Although the NATO alliance
regularly threatened to initiate tactical nuclear use in Europe
should the Warsaw Pact attack, a major disincentive was always
the catastrophic damage to friendly civilians which would result
from the scale of battlefield nuclear use necessary to offset militarily
NATO's conventional inferiority.[36]

Historically, psychological arguments have been at least as
influential as straightforward considerations of cost-effectiveness
in discussions about resorting to nuclear use. In 1945 the United
States was already destroying Japanese cities one after another
with conventional bombing raids, and the main argument for
using atomic weapons instead was that the psychological shock
caused by this new form of attack might finally impel the Japanese
to give in.[37] During the Cold War, NATO leaders stressed the
political as much as the military utility of first nuclear use, hoping
that the fear of escalation produced by a nuclear 'demonstration'
would persuade the Soviet Union to cease its attack and withdraw.[38]
Had NBC capabilities as a whole possessed clearer and more
tractable *military* utility in a wider variety of tactical circumstances, their use over the past eighty years would almost certainly
have been less restrained.

The restraining influence of dubious military utility has often
been compounded by logistical difficulties. Chemical weapons have
once again been the worst affected, thanks to the inconvenience
and danger of producing and deploying large quantities of noxious
chemicals and the associated protective equipment for one's own
forces. This was not such a problem in the First World War or the
Iran–Iraq War, thanks to the generally static nature of the battle-
fronts, but it has imposed much greater restraints during mobile
warfare.

General Ochsner argued that German use of chemical weapons
in the invasion of the Soviet Union in the Second World War

would have strained to the breaking point our supply service, which was
difficult enough anyhow in view of the poor railroad communications, the
inadequacy of roads for modern motor transport, and the great distance

[35] See Betts, *Nuclear Blackmail and Nuclear Balance*, 39, 51.
[36] See M. Legge, *Theater Nuclear Forces and the NATO Strategy of Flexible
Response* (Santa Monica, Calif.: RAND report R-2964FF, 1983).
[37] See Freedman, 'The Strategy of Hiroshima'.
[38] See Legge, *Theater Nuclear Forces*.

from the German bases. We had to do everything possible to avoid this happening.[39]

Logistical difficulties also affect *defending* states when faced with a fast-moving armoured offensive, and the problem of shipping the requisite supplies to the right place in the face of enemy air interdiction may help to explain why France in 1940, the Soviet Union in 1941, and Iraq in 1991 did not initiate the use of gas in their defence.

In 1945 logistical problems were a major reason why the United States did not initiate chemical warfare against the now defenceless Japanese. Insufficient gas munitions were available in forward areas, and the Joint Chiefs of Staff decided that the shipment of conventional supplies for the invasion of Japan should take priority.[40] There were no similar logistical difficulties in moving forward the small *atomic* stockpile then available, which helps to explain why the entire stockpile was used against Japan at the earliest opportunity.

The fact that so few atomic bombs were available in 1945 acted not as a restraint but as a spur to escalation. Proposals for attacking a military target or conducting a demonstration shot so as to minimize civilian casualties were rejected, partly on the grounds that the few weapons available had to be used in the most effective and shocking manner.[41] However, stockpile shortages did have a restraining effect five years later during the Korean War, when the tactical use of atomic weapons was resisted partly because it would waste the limited stockpile then available, thus leaving the United States ill prepared for a possible war in Europe.[42]

The initial spur to the Allied development of atomic weapons during the Second World War was as a hedge against a possible German breakthrough in this area. However, the project soon acquired its own momentum, and one reason why there was so much pressure towards the eventual use of the bombs against Japan was that some direct wartime return had to be demonstrated

[39] Ochsner, *History of German Chemical Warfare*, 21.
[40] See F. Brown, *Chemical Warfare: A Study in Restraints* (Princeton, NJ: Princeton University Press, 1968), ch. 6.
[41] See Freedman, 'The Strategy of Hiroshima'.
[42] Betts, *Nuclear Blackmail and Nuclear Balance*, 36–7.

for the enormous resources invested in the Manhattan Project.[43] This highlights an important general point about NBC readiness— namely that, if large investments are made in offensive and defensive preparations to deter *enemy* initiation of warfare, there may be a temptation to initiate such warfare oneself if other restraints should disappear.[44]

What are the implications of all of this for NBC restraints in the Middle East? As in the past, chemical weapons seem most likely to be used in drawn-out static conflicts, or against unprotected tribesmen like the Kurds. Nuclear weapons, since so few will be available (except perhaps to Israel), are unlikely to be used in tactical operations, but will be employed for coercive purposes or held in reserve as a counter-city deterrent. NBC use will probably continue to be avoided in swift mobile campaigns like the 1991 Gulf War or the various Arab–Israeli wars. The main danger is that paranoia about enemy NBC capabilities may lead to an arms race in such weapons and in protective countermeasures, with the states involved then being tempted to make use of these capabilities against opponents such as dissident groups or weaker countries who are not so well equipped.

### CULTURAL AVERSION

Even when all the foregoing restraints are taken into account, there remain plenty of instances in which states have held back from using NBC weapons during a conflict, for no obvious military reason and with no fear of retaliation to constrain them. During the Falklands War, for example, the United Kingdom engaged in a very risky conventional campaign, without any thought of using its nuclear arsenal to coerce Argentina or to strike the mainland airbases being used to attack the UK fleet.[45] The explanation for such restraint must be sought primarily in cultural and political considerations.

Throughout history, various weapons have been viewed with opprobrium in particular cultures because they were perceived

---

[43] See L. Giovannitti and F. Freed, *The Decision to Drop the Bomb* (London: Methuen, 1967).
[44] See General Marshall's comments in Brown, *Chemical Warfare*, 273–4.
[45] See L. Freedman and V. Gamba-Stonehouse, *Signals of War* (London: Faber & Faber, 1990).

as 'unfair'.[46] Chemical and biological weapons have attracted especially widespread and enduring odium, probably because of their insidious character. Nuclear weapons came to be seen as similarly repulsive, as awareness spread during the 1950s of the effects of fall-out and radiation poisoning, with its potential impact upon generations yet unborn.[47] The disproportionate vulnerability of civilian populations to NBC weapons has also contributed to the widespread distaste in which such weapons have been held, due to the prevailing belief that it is more legitimate to harm combatants than non-combatants.[48]

What is particularly interesting about cultural aversion to NBC weapons is that it has not so far followed the pattern seen with other 'unfair' weapons ranging from crossbows to air-delivered bombs—namely, one of gradual acceptance of the weapon as a normal component of warfare. Instead, both chemical and nuclear weapons were employed most freely at the time when they were first developed, in the First and Second World Wars respectively. Thereafter, inhibitions grew, and states often held back from employing NBC weapons, even without compelling military or political reasons for such self-restraint.

Nor has NBC use followed the historical pattern that cultural aversion to the use of particular weapons is strongest between culturally similar antagonists. Although Italian actions in Somalia and US use of atomic bombs and irritant gases against Asian rather than European opponents do fit into this pattern, most other instances of chemical employment over the past eighty years have taken place between near neighbours and co-religionists. A more important factor is that non-Western societies such as Japan and Iraq have not necessarily displayed the same moral qualms about the use of chemical and biological weapons as have developed in Western cultures.[49]

This difference may have been due in part to the influence of public opinion in liberal democratic societies. US use of irritant agents and defoliants in Vietnam prompted an outcry at home,

[46] See M. van Creveld, *Technology and War: From 2000 BC to the Present* (New York: Free Press, 1989), 70–3.
[47] See N. Moss, *Men Who Play God: The Story of the Hydrogen Bomb* (London: Gollancz, 1968), 84–100.
[48] See M. Howard (ed.), *Restraints on War: Studies in the Limitation of Armed Conflict* (Oxford: Oxford University Press, 1979).
[49] See Brown, *Chemical Warfare*, 246–9.

and, although this outcry died down over time, it helps to explain why chemical operations never escalated to the use of lethal agents as in other conflicts.[50] However, democratic publics have by no means always opposed the use of NBC weapons. There was no significant public protest over the escalation of chemical hostilities in the First World War, and US Press articles and public opinion polls towards the end of the Second World War indicate an increasing readiness to *initiate* the use of gas in the bitter island warfare against the Japanese.[51] Although the atomic bombing of Hiroshima and Nagasaki in 1945 aroused many second thoughts in the United Kingdom and the United States in later years, it was hailed at the time as a welcome alternative to a costly conventional invasion of Japan.[52]

An equally important determinant of restraint, in non-democratic as well as democratic states, has been the attitude of individual leaders. It was Falkenhayn's insistence on the use of chlorine in 1915 which ensured that lethal-gas warfare was initiated, despite the many reservations of German field commanders.[53] Hitler, on the other hand, displayed considerable personal aversion to gas warfare, after being temporarily blinded at Ypres in 1918. This initially inhibited German consideration of chemical warfare, though it did not stop the use of gas in the death camps and it may at the last have encouraged Hitler to order chemical use on the battlefield precisely because of its horrific character.[54] Roosevelt's hatred of gas was a definite restraint on US initiation of chemical warfare, and it was not until his replacement by Truman that US use of gas against Japan could be considered seriously.[55]

Perhaps the most important restraining influence of cultural aversion to NBC weapons has been its role in hindering the assimilation of such weapons by armed forces themselves. Military dislike of gas was evident even in the First World War, with constant references to its 'unchivalrous' character and to the dangers which it posed to friendly troops.[56] After the war, the

[50] See Adams, *Chemical Warfare*, 80–4.
[51] See Brown, *Chemical Warfare*, 287–8.
[52] See R. Miles, 'Hiroshima: The Strange Myth of Half a Million American Lives Saved', *International Security* (fall, 1985), 121–40.
[53] See Trumpener, 'The Road to Ypres'.
[54] Brown, *Chemical Warfare*, 235–8.
[55] SIPRI, *Problem of Chemical and Biological Warfare*, i. 316–17.
[56] See W. Moore, *Gas Attack!* (London: Leo Cooper, 1987).

various Chemical Corps tried to overcome this resistance, but generally failed to convince the other military arms, fearful for their established roles. Although the fact that chemical weapons wounded more than they killed did lead some individuals such as Churchill to argue that 'Gas is a more merciful weapon than high explosive shell', most military men remained viscerally opposed to the use of poison in war.[57]

The failure of chemical weapons to win acceptance as 'normal' military instruments seems to have played a major role in their non-use in the Second World War. Frederic Brown argues that, in the case of the United Kingdom and Nazi Germany, 'The inhibition of non-assimilation within the military was at least equal in importance to possession of a deterrent military capability as a restraint on initiation.'[58] Although US military attitudes were more open-minded, Brown suggests that, 'The combination of personal and institutional dislike of toxic agents and the estimated marginal overall effectiveness of gas as a weapons system were insurmountable restraints on employment.'[59] Julian Perry Robinson concludes similarly that, 'The reluctance to accept gas as a useful weapon stemmed at least as much—probably far more—from institutional pressures and psychological constraints as from rational consideration of its military utility.'[60]

Biological weapons have encountered even greater prejudice from military men, thanks to their mode of operation and their intractability as tactical instruments. Nuclear weapons have won more converts, especially among air forces, but armies and navies have tended to see them as a threat to their conventional military roles, and so to emphasize more traditional ways of war.[61] The result has been that military commanders have very rarely taken the lead in pressing for the use of NBC weapons, and have often reacted with scepticism or inertia when such use has been suggested to them. Had military attitudes been different, many of the practical and logistical obstacles to NBC employment might have been easier to overcome.

Cultural aversion to NBC weapons seems unlikely to prove

[57] See Adams, *Chemical Warfare*, 46–9.
[58] Brown, *Chemical Warfare*, 246.
[59] Ibid. 282.
[60] SIPRI, *Problem of Chemical and Biological Warfare*, i. 334.
[61] See, e.g., M. Taylor, *The Uncertain Trumpet* (New York: Harper & Row, 1960).

sufficient in itself to restrain Middle Eastern states from resorting to such weapons in the future, given the lack of domestic outcry over Egyptian and Iraqi use of poison gas in recent decades. However, the qualms of certain leaders and of Israeli public opinion may be expected to have some restraining effect, as will the fact that the armed forces of the region seem no keener to become reliant on NBC use than their counterparts elsewhere. Where cultural aversion *will* have a major impact is on the conduct of Western intervention forces. The 1991 Gulf War illustrated the reluctance of Western states to consider NBC use, *even in retaliation*, against ordinary Iraqis thrust into war by Saddam Hussein. How to deter NBC attack in the face of such perceived distinctions between ruler and ruled will be a major challenge for Western powers in the future.

POLITICAL CONSIDERATIONS

A final restraining influence on NBC use has been anticipation of an adverse political reaction on the part of neutral or allied states. US initiation of chemical warfare at the end of the Second World War was hindered by the vulnerability of allies such as the United Kingdom to Axis retaliation—even China opposed such a US initiative, lest Japan attack Chinese cities in response.[62] Similar constraints inhibited US use of *nuclear* weapons in the early years of the Cold War; in 1954 Dulles reported the British to be 'scared to death' and 'almost pathological in their fear of the H-bomb'.[63] It is not only the senior partner in an alliance which may be influenced in this way—a major constraint on Israeli use of its NBC capabilities has undoubtedly been fear of the reaction of the United States.

The restraining influence of world opinion more generally is harder to evaluate. NBC weapons have certainly acquired considerable stigma within the international community, and decision-makers have often expressed fears of adverse world reaction during debates over NBC initiation. This restraint seems to have had greatest force when countries were concerned to avoid

[62] See Brown, *Chemical Warfare*, 278–81.
[63] See Betts, *Nuclear Blackmail and Nuclear Balance*, 52.

alienating client or non-aligned states (as during the Cold War), or when they were dependent on outside support and wished to avoid punitive sanctions.

Concern about world opinion is illustrated by the lengths to which states have gone to conceal their use of NBC weapons, or to shift the blame to the other side. Often this tactic has been successful, as in the long-running debate over whether chemical weapons were used in Kampuchea, Laos, and Afghanistan. However, it is hard to maintain such outside doubts while using NBC weapons in an unrestrained fashion, so the desire for secrecy has imposed its own constraints on NBC use. Sensitivity to world opinion also helps to explain why chemical warfare has often begun with irritant agents. When international reaction has been muted, as in the case of the Iran–Iraq War, escalation to lethal gases has generally occurred.

Formal international agreements have a mixed record in restraining the use of NBC weapons. Participants at the Hague Peace Conferences in 1899 and 1907 committed themselves 'to abstain from the use of projectiles, the sole object of which is the diffusion of asphyxiating or deleterious gases', but this did not stop nations from experimenting with such devices, and actually using irritant gases when war came.[64] Germany defended its use of irritant gas shells on the specious grounds that they contained explosive as well as gas, and it justified its later use of chlorine because it was released from cylinders rather than projectiles.[65]

The Geneva Protocol of 1925 prohibited 'the use in war of asphyxiating, poisonous or other gases and of bacteriological methods of warfare' between the signatory nations.[66] It did not stop these nations making chemical and biological preparations and reserving the right to retaliate if attacked, nor did it dissuade Italy from using gas in Ethiopia 'in punishment of such abominable atrocities as those committed by the Ethiopian forces'.[67] However, it seems to have had some indirect impact on the non-use of gas in the Second World War. In Julian Perry Robinson's words,

[64] See L. F. Haber, *The Poisonous Cloud* (Oxford: Oxford University Press, 1986), 18–25.
[65] Trumpener, 'The Road to Ypres', 468–9.
[66] See SIPRI, *Problem of Chemical and Biological Warfare*, iv. *CB Disarmament Negotiations, 1920–1970*, ch. 2.
[67] Ibid. 180.

The legal constraints were important...not because of any direct influence on the decision whether or not to initiate chemical warfare, but rather because of their influence in retarding acceptance of gas as a standard weapon of war, and hence in their contribution to the belligerents' overall unpreparedness to wage chemical warfare, and their leaders' unwillingness to authorize it.[68]

In terms of restraining influence, unilateral statements by individual states seem to have had just as much impact as international agreements. The United States in 1945 was constrained in its consideration of chemical warfare by Roosevelt's clear pronouncement two years earlier than, 'we shall under no circumstances resort to the use of such weapons unless they are first used by our enemies'.[69] The Soviet Union under Brezhnev made a similar pledge that it would never use nuclear weapons first, and the nuclear powers of NATO, although reserving the right of first nuclear use against an overwhelming Warsaw Pact attack, pledged in 1978 that they would never initiate nuclear war against a non-nuclear state not in alliance with a nuclear power.[70]

As the norm of non-use of nuclear weapons has become established over time, it has come to exercise a powerful restraint in its own right upon states considering breaking the taboo. Conversely, the employment of chemical weapons in Vietnam, Yemen, the Persian Gulf, and elsewhere has eroded the taboo on gas warfare which developed during the Second World War, making it easier to breach this threshold again in the future.

The decline of close client-state relationships now that the Cold War is over is likely to make Middle Eastern states considering NBC use less concerned about offending their great power allies. This makes it more important that such states should be dissuaded by fears of adverse political reactions within the international community as a whole. The challenge for the great powers is to exploit the opportunity offered by the end of the Cold War for a concerted stand against NBC employment, without triggering further NBC proliferation among Middle Eastern states concerned to escape domination by a great power condominium.

[68] SIPRI, *Problem of Chemical and Biological Warfare*, i. 320.
[69] See Brown, *Chemical Warfare*, 264.
[70] On the problems of a more clear-cut declaration, see J. Steinbruner and L. Sigal (eds.), *Alliance Security: NATO and the No-First-Use Question* (Washington DC: Brookings, 1983).

## CONCLUSION

When viewed in the light of historical experience, the fact that chemical weapons were not employed during the 1991 Gulf War appears far less surprising. Iraq's use of such weapons against Iran took place in a static conflict against ill-protected troops, with Iraqi air and artillery superiority, with no immediate prospect of Iranian retaliation, and with many outside powers tacitly taking Iraq's side, due to their fears of Iranian fundamentalism. Even then, Iraq limited its use of gas to critical tactical situations, and did not employ it against Iranian civilian targets.[71] Small wonder that, in 1991, Iraq was neither willing nor easily able to use chemicals in a mobile war against well-protected forces, having lost command of the air, and facing the prospect of devastating retaliation either in kind or through an expansion of coalition war objectives to include the overthrow of the Ba'ath regime.[72]

This pattern of NBC use and restraint fits squarely with previous historical experience, and suggests that, even if NBC capabilities continue to spread within the Middle East, their actual employment in time of conflict is likely to remain severely circumscribed. Arguments that maverick leaders such as Saddam Hussein or Colonel Gaddafi will employ NBC weapons at the first opportunity are strikingly similar to earlier misplaced fears about the acquisition of nuclear weapons by communist China, and neglect historical counter-arguments, such as the fact that Hitler's Germany refrained from using nerve gas even when it was being overrun.

This having been said, it would be wrong to suggest that the spread of NBC capabilities in the Middle East would be a good thing, bringing to that region something of the deterrent stability which has prevailed among the great powers over the past forty years.[73] Political instabilities and hatreds in the Middle East are deeper; hostile relationships are multipolar rather than bipolar; and an unbridled emphasis on the building-up of deterrent NBC capabilities risks such weapons falling into the hands of terrorists or being assimilated by the various military forces to the point where they are used whenever militarily advantageous.

[71] See Efraim Karsh, Ch. 2, this volume.
[72] See Martin Navias, Ch. 3, this volume.
[73] The classic articulation of this minority view is K. Waltz, *The Spread of Nuclear Weapons: More May be Better* (Adelphi Papers, 171; London: IISS, 1981).

In this situation, the ideal outcome remains NBC disarmament as part of the Middle East peace process or through wider global initiatives such as the CWC. Should certain NBC capabilities persist, a major priority is to avoid circumstances in which the use of those capabilities becomes a relatively cost-free option, as it was for Egypt in Yemen and for Iraq against the Iranians and Kurds. Since further proliferation of national NBC deterrent forces is dangerous and does not address the problem of conflicts *within* Middle Eastern states, it would be preferable for the great powers to strengthen the *political* taboo against NBC employment, by posing a clearer threat of sanctions against states which initiate such action, whatever the political circumstances. Such a policy contains enormous problems, but should help to halt and perhaps reverse NBC proliferation in the Middle East as well as mitigating the consequences if NBC capabilities persist.

# 2

# Rational Ruthlessness: Non-Conventional and Missile Warfare in the Iran–Iraq War

## EFRAIM KARSH

As Iraqi missiles pounded Israeli and Saudi Arabian cities during the 1991 Gulf War, the fear among many in the region and beyond was that Saddam Hussein would escalate this campaign and use chemical weapons against coalition forces and Israel. Given the horrendous role of gas in twentieth-century Jewish history, there was wide unanimity that any chemical attack against Israel would be met with harsh response, perhaps even a nuclear strike against Iraq.

These fears are not difficult to understand. The common perception of Saddam Hussein at the time was of an irrational and impetuous megalomaniac who, like the German dictator Adolf Hitler, would stop at nothing in his determined drive for regional mastery. Saddam himself sought to cultivate this stark image throughout the crisis by threatening the anti-Iraq coalition with 'the mother of all battles'. More specifically, he capitalized on his notorious resort to chemical weapons during the Iran–Iraq War and implied the possible repetition of this option in a future war by openly loading and unloading chemical bombs on Iraqi fighting aircraft.

And yet, to the extent that past conduct provides an indication of future behaviour, Saddam's threats should not have been taken at face value. His outspoken determination to hurl missiles at Israel seemed serious enough, given the extensive use of surface-to-surface missiles against civilian targets during the Iran–Iraq War. His implicit threat to use chemical weapons in the event of a coalition attack against Iraq, on the other hand, was not substantiated by Iraq's chemical warfare during Iran–Iraq War.

Though devoid of any moral inhibitions, as shown by the indiscriminate gassing of hapless civilians in Kurdistan, Iraq's battlefield use of chemical agents tended to be incremental and heavily circumscribed, taking place only where there was absolutely no risk of retaliation.

In this chapter I will examine the record of non-conventional warfare during the Iran–Iraq War. Focusing mainly on Iraq, which was by far the main employer of non-conventional means, I will argue three points. First, for all its ferocity, non-conventional escalation was a deliberate and rational process, subject to cost–benefit calculations and susceptible to deterrence, having thresholds and limitations. Secondly, the overall impact of non-conventional warfare on the general course of the war was marginal, though it was of tactical significance in several battles. Non-conventional means proved most effective when used either in conjunction with other means or in circumstances in which opposing forces had long been debilitated and demoralized. Thirdly, in the Iraqi case this relative circumspection reflected the cautious nature of Saddam Hussein and his limited war objectives. In the case of Iran, which was more prone to escalation, it stemmed from dire material and technical constraints.

## THE EVOLUTION OF THE BELLIGERENTS' NON-CONVENTIONAL ARSENAL

Iraq's chemical (and biological) programme reportedly dates back to 1974, when a three-man committee headed by Saddam Hussein, then Vice-President, was formed for this purpose.[1] The committee soon established contact with a Beirut-based consulting firm by the name of Arab Projects and Development, owned by two Palestinian construction tycoons. At the firm's advice, and through its good offices, Saddam began recruiting Arab scientists and technicians from all over the world. Between 1974 and 1977 more than four thousand research personnel had been lured into

[1] The other members of the committee were Adnan Khairallah Talfah, Saddam's cousin and later Iraq's Minister of Defence, and Adnan Hussein al-Hamdani, member of the Revolutionary Command Council, Iraq's supreme ruling body. Both were to die of unnatural causes: Hamdani was executed in 1979 for participation in an alleged pro-Syrian attempt to overthrow the Iraqi regime, and Talfah died in a mysterious helicopter crash in May 1989.

Iraq by substantial financial rewards, and assigned the task of constructing chemical and biological plants.

Since this Arab scientific effort still failed to make Iraq self-sufficient in chemical warfare, Saddam decided in the late 1970s to seek the support of foreign companies for his chemical and biological programmes. Procurement teams were sent to Europe and the United States, again disguised as commercial representatives of Arab Projects and Development, to search for the necessary technological know-how and supplies. Before long the Iraqis established contacts with a US company which provided the blueprints for the construction of the first Iraqi chemical-weapons plant. The schemes were labelled as 'flow charts for a pesticide plant', but 'even a novice would have recognized that at least two of the chemicals it was supposed to produce, Amaton and Paratheon, could be used to make nerve gas'.[2] Although eventually the US company did not receive the necessary licence to export the machinery for constructing the chemical plant, Saddam had managed to obtain a key element for his project: the plans of the enterprise.

The Iraqi efforts in Europe were not much more successful. Several companies in the United Kingdom, Germany, and Italy were approached with the request to assist Iraq in assembling a chemical plant, but to no avail. On at least one occasion—involving the large UK chemical corporation, Imperial Chemical Industries—the Iraqi agents actually aroused suspicion, which, in turn, drove the company to inform the Foreign Office on the nature of the Iraqi interest. Although the UK authorities failed to take any action, Hussein was deterred by the incident and decided that Iraq would assemble the plant on its own. Subsequently, Baghdad began purchasing the required components on a piecemeal basis under the guise of constructing a pesticide plant, and by mid-1979 had completed the assembly of its first chemical-warfare plant near the north-western town of Akashat.

By the outbreak of the Iran–Iraq War, then, Iraq's chemical stockpiles were rather limited, and its production of chemical-warfare agents was minimal. It was only after being forced on to the defensive in 1982 that Saddam embarked on his ambitious

---

[2] *Independent* (London), 12 Sept. 1990.

programme of chemical armament.[3] Within less than a decade, with the support of numerous foreign companies, German in particular, he was to transform Iraq into a major chemical power, in possession of some forty-six thousand chemical weapons ranging from missile warheads to artillery shells to aerial bombs.[4] These means of delivery were armed with home-made gases including a refined form of distilled mustard (HD), as well as tabun nerve agent and the more potent VX nerve agent.[5]

The evolution of Iraq's ballistic-missile programme was similar. From a humble beginning of twenty-six dated Frog-7 and 12 Scud-B launchers (with a range of approximately seventy and three hundred kilometres respectively) in September 1980,[6] it developed into the largest missile force in the Middle East, boasting a few hundred missile launchers and more than a thousand missiles. These extensive stockpiles included an unknown number of Iraqi-upgraded Scud missiles: the Al-Hussein with a range of about six hundred kilometres and the Al-Abbas with a range of about nine hundred kilometres.[7]

The rationale behind Iraq's non-conventional-arms industry was not purely military. For Saddam, weapons of mass destruction (including nuclear arms) had never been yet another category of armament; they were a key to political survival. Ballistic missiles and nuclear weapons would blunt Israel's military and technological edge and banish any aggressive designs on the part of Iraq's hostile neighbours, Iran and Syria in particular. Chemical weapons, on top of their pure military value, were the ultimate weapons against domestic opponents, as the Kurds were to learn in the most horrendous fashion.

Iran's non-conventional-arms industry was far less ambitious, and evolved, by and large, as a response to Iraq's formidable arsenal. Before the war Iran had had no chemical-weapons programme, and its interest in the development of indigenous production capability emerged only in 1983–4, following Iraq's growing

---

[3] A. H. Cordesman and A. R. Wagner, *The Lessons of Modern War*, ii. *The Iran–Iraq War* (Boulder, Colo.: Westview & Mansell, 1990), 510.

[4] *The Times* (London), 31 July 1991.

[5] *Sunday Times* (London), 2 Sept. 1990.

[6] IISS, *The Military Balance, 1980–1981* (London, 1980), 42.

[7] Al-Abbas was reportedly tested in the late 1980s but was not yet operational by the outbreak of the Kuwaiti crisis in 1990. See M. S. Navias, 'Ballistic Missile Proliferation in the Middle East', *Survival* (May–June 1989), 225–41.

use of gas on the battlefield. The road to the attainment of this goal was very much the same as Iraq's—namely, approaches to Western companies (again the Germans figured prominently among Iran's suppliers) for support in establishing 'pesticide plants'. In 1986–7 Iran was believed to produce sufficient lethal agents for its means of delivery (aerial bombs and artillery shells), and by the end of the war in July 1988 it was reportedly producing significant amounts of mustard and nerve gas.[8]

Iran's procurement of ballistic missiles underwent a similar process. Internationally isolated and severed from its main arms suppliers following the Islamic Revolution, Tehran sought to buy surface-to-surface missiles on the 'black market' and from its few remaining allies—Libya, Syria, and North Korea. These efforts were crowned with success, and in 1985 Iran's modest missile force reportedly made its baptism of fire. Simultaneously, and because of its allies' limited capacity for military support, Iran launched its own production programme: it improved the Scud-B missile and manufactured both an indigenous short-range missile, the Oghab (up to forty kilometres), and a 130-kilometre missile, the Iran-130.[9]

### CHEMICAL WARFARE IN THE WAR

When exactly the first battlefield use of chemical weapons occurred is difficult to say. According to Iranian allegations, made in the initial stage of the war, Iraq used weapons that 'spread germs' during the battle for the city of Sussangerd in November 1980.[10] This claim was widely dismissed at the time as a propaganda ploy, and the Iranians themselves later retracted from their early allegations. An official Iranian report in 1984 traced the first Iraqi use of mustard gas to October 1982, while, in a comprehensive report submitted to the United Nations in 1988, Tehran argued that Iraq had already resorted to chemical warfare in January 1981.[11]

To my best judgement, Iraq's first chemical experience took

---

[8] Cordesman and Wagner, *Lessons of Modern War*, 513.
[9] Ibid. 498; Navias, 'Ballistic Missile Proliferation', 229.
[10] *Washington Post*, 17 Nov. 1980.
[11] Conference on Disarmament, Letter dated 11 April 1988 from the permanent representative of the Islamic Republic of Iran addressed to the President of the Conference on Disarmament, *CD/827*, 12 Apr. 1988.

place much later, in July 1982, in the course of a large-scale Iranian thrust in the direction of Basra, Iraq's second largest city. The actual use of gas was extremely circumspect: Iraq did not go beyond the employment of non-lethal tear gas in a small segment of the battlefield, and resorted to this action only after warning the Iranians in advance that there was a 'certain insecticide for every kind of insect'.[12] Views differ as to how effective this first use was. While some accounts describe the episode as counter-productive for the Iraqis—a change in wind direction reportedly blew the tear gas back to them—others view it as a major success: the tear gas was said to have frustrated the operations of an entire Iranian division.[13]

Be that as it may, the fact that Saddam had forgone the use of chemical weapons until that relatively late stage in the war, when Iraqi forces were already defending their own territory against Iran's successive human-wave attacks, and even then chose to use non-lethal chemical agents and to warn the Iranians in advance, indicated his strong interest in limiting the war and in avoiding uncontrollable escalation. From the beginning of the war he had sought to confine the war by restricting his army's goals, means, and targets. Thus, his territorial goals did not go beyond the Shatt al-Arab and a small portion of Khuzestan. As to means, the invasion was carried out by less than half of the Iraqi army—five of twelve divisions—and the invading force did not use non-conventional means, such as chemical agents. As to targets, Baghdad's initial strategy was one of counter-force, avoiding targets of civilian and economic value and in favour of attacks almost exclusively on military targets.

Saddam Hussein probably hoped that a limited campaign would persuade Iran's revolutionary regime to desist from undermining his personal rule.[14] By exercising self-restraint, he sought to signal an intent to avoid a general war, with the hope that Tehran would

---

[12] *Newsweek*, 2 Aug. 1982. On Iraq's first use of gas, see E. Karsh, *The Iran–Iraq War: A Military Analysis* (Adelphi Papers, 220; London: IISS, 1987), 26; T. L. MacNaugher, 'Ballistic Missiles and Chemical Weapons: The Legacy of the Iran–Iraq War', *International Security*, 15/2 (fall, 1990), 17; A. Terrill, 'Chemical Weapons in the Gulf War', *Strategic Review* (spring 1986), 51–8; Cordesman and Wagner, *Lessons of Modern War*, 514.

[13] See, e.g., *Washington Post*, 5 Apr. 1988; *Newsweek*, 2 Aug. 1982.

[14] On Saddam's decision to invade Iran, see E. Karsh and I. Rautsi, *Saddam Hussein: A Political Biography* (New York: Free Press, 1991), ch. 6.

respond in kind, and perhaps even be willing to reach a settlement. In pursuit of this goal, Saddam voluntarily halted the advance of his troops within a week after the onset of hostilities, and then announced his willingness to negotiate a peaceful settlement.

As is well known, the mullahs in Tehran did not live up to Saddam's expectations. Instead of seeking a quick accommodation, they capitalized on his gesture, which turned out to be a strategic miscalculation of the highest degree, in order to seize the initiative and to move to the offensive. By mid-1982 they had already expelled Iraqi forces from their territory and begun persistent thrusts into Iraq. As the fearful spectre of defeat loomed larger and larger, Saddam increasingly fell back on the use of chemical weapons.

The first confirmed use of mustard gas occurred in October–November 1983, in the course of a large Iranian offensive in the northern region of Kurdistan (Operation Dawn IV).[15] It was subsequently used in February and March 1984, during a series of interconnected Iranian offensives (Dawn V, Dawn VI, and Khaybar). These offensives were the largest engagement in the war until then, pitting some five hundred thousand men under arms against each other along a hundred-mile front. On several occasions the Iranians nearly breached Iraq's formidable line of defence. They captured the Majnoun Islands on the southern front and came within fifteen miles of the Baghdad–Basra highway, only to be contained with great efforts and through the use of chemical weapons.[16]

A year later, in March 1985, the Iranians resumed their efforts to reach the Baghdad–Basra highway and thus cut Iraq into two. Having abandoned their frontal human-wave assaults in favour of more conventional warfare, carried out under the leadership of the professional army, they came closer than ever to their goal. Inflicting heavy casualties on the Iraqis (reportedly between ten and twelve thousand, compared with Iran's loss of some fifteen

[15] Cordesman and Wagner believe that mustard gas was used for the first time in December 1982, in the course of a large-scale Iranian offensive in the direction of Basra (*Lessons of Modern War*, 508, 510). MacNaugher, conversely, thinks such use occurred only in early 1984 ('Ballistic Missiles'), 17.

[16] Cordesman and Wagner, *Lessons of Modern War*, 514–16; E. O'Ballance, *The Gulf War* (London: Brassey's, 1988), 116, 118, 140.

thousand), they managed briefly to cut part of the Baghdad–Basra highway nearest the border. The shaken Saddam responded by ordering the widest use of chemical weapons until then. By this time the Iraqis had already broadened their use of chemical agents beyond mustard gas to include such agents as tabun and cyanide.[17]

The Iraqi *modus operandi* in all these cases was identical. Though widening the use of gas by the offensive, Iraq resorted to chemical warfare only after its threats in this respect had been ignored by the Iranians, and only at critical moments, when there was no other way to check Iranian offensives. 'If you execute the orders of Khomeini's regime,' cautioned an Iraqi High Command statement in September 1983, 'Iraq would use, for the first time in the war, modern weapons that had not been used in previous attacks for humanitarian reasons... [these weapons] would destroy any moving creature on the front.'[18] Ignorance of this warning, aimed at forestalling the launching of Operation Dawn IV, resulted in the first use of lethal gas in the war.

It was only in 1986 that Iraq cautiously moved to what may be termed as *offensive* employment of gas. In February of that year the Iranians managed to gain their most significant foothold on Iraqi territory with the occupation of the Fao Peninsula at the southern tip of Iraq. The Iraqis responded by mounting a major counter-offensive under the leadership of Maher Abd al-Rashid, Saddam's kinsman and his son's father-in-law, and one of Iraq's best performing military commanders during the war. Although the counter-offensive was carried out by Iraq's élite units, the Republican Guard, and even though the Iraqis enjoyed over-whelming superiority in firepower and resorted to chemical weapons, the Iraqi effort failed shamefully, with some ten thousand Iraqis killed in one week.[19] In these circumstances, Saddam Hussein was understandingly anxious to achieve some visible successes. In mid-May 1986, in a massively publicized operation, Iraq took the Iranian town of Mehran on the central front (again with the use of gas) and offered to exchange it for Fao. The offer was spurned

---

[17] Karsh, *The Iran–Iraq War*, 27–31; O'Ballance, *The Gulf War*, 161–2.

[18] S. Chubin and C. Tripp, *Iran and Iraq at War* (London: Tauris, 1988), 59.

[19] Four years later, during his famous meeting with the US Ambassador to Baghdad, Ms April Glaspie, before Iraq's invasion of Kuwait, Saddam turned this humiliating defeat into a shining achievement. 'Yours is a nation that cannot afford to lose 10,000 men in one battle,' he boasted in front of the startled Ambassador.

and Iranian forces recaptured Mehran in early July. Three months later the Iranians managed to drive a few miles into Iraqi territory in Kurdistan.

Following this string of military setbacks, Saddam was confronted, for the first and only time in his career, by what nearly amounted to an open mutiny. With the Iranian army at the gates of Basra, the military leadership rose up in an attempt to force Saddam to win the war despite himself. Fortunately for the Iraqi President, the generals did not demand political power nor try to overthrow him. All they wanted was the professional freedom to run the war according to their best judgement, with minimal interference from the political authorities. Caught between the hammer and the anvil, Saddam gave in to his generals and the war gradually took a positive turn. In early 1987 the Iraqi army rebuffed Iran's last major offensives during the war. A year later it moved to the offensive and, in a series of shining successes, brought Tehran to its knees and forced it to beg for a ceasefire.

In all these offensives Iraq made use of chemical weapons. Gas was employed during the Iraqi counter-offensive in Kurdistan in March 1988, in the recapture of Fao in mid-April, the liberation of the Majnoun Islands, and the incursion into Iran on the central front two months later.[20]

Apart from Saddam's determination to get the Iranians off Iraqi territory at all costs, Iraq's growing resort to chemical warfare in offensive operations reflected the generals' more lax attitude towards this operational mode. For all his lack of moral inhibitions and respect for international norms, Saddam's overwhelming preoccupation with his political survival injected a strong element of restraint into his behaviour, which his generals lacked completely. For them chemical weapons were yet another category of armament, and their use depended on their pure military value in the relevant circumstances. As Maher Abd al-Rashid put it: 'If you gave me a pesticide to throw at these swarms of insects to make them breathe and become exterminated, I would use it.'[21]

No one knew this better than the Kurds. In 1987 and 1988 they were subjected to a brutal punitive campaign involving the extensive use of chemical weapons, including mustard gas, cyanide, and

[20] See, e.g., *Washington Post*, 24 Mar., 4 Apr. 1988.
[21] *Newsweek*, 19 Mar. 1984.

tabun nerve agent against unprotected civilian population. The
first attacks of this kind were reported in May 1987, when some
twenty Kurdish villages were gassed in an attempt to deter the
civilian population in Kurdistan from collaborating with the
advancing Iranian forces. A month later several Kurdish villages in
Iran were given the same 'medicine', with some hundred people
dead and two thousand injured. The most appalling attack
took place in March 1988, when the spectre of a major Iranian
breakthrough in Kurdistan drove the Iraqis to employ gas on an
unprecedented scale against the Kurdish town of Halabja. As the
thick cloud of gas spread by the Iraqi planes evaporated into
the clear sky, five thousand people—men, women, children, and
babies—were left dead. Nearly ten thousand suffered injuries.[22]

Iran's use of chemical warfare was a far cry from that of Iraq.
As noted earlier, Tehran's chemical arsenal during the first years
of the war was virtually non-existent and its first use of gas in the
war, which reportedly took place in 1984 or in 1985, was limited
to firing unexploded Iraqi artillery and mortar shells.[23] Others set
the date of the first documented Iranian employment as late as
1988.[24] Either way, there is general agreement that Iran's resort to
chemical warfare was sporadic and extremely limited.

Furthermore, the Iranians were ill equipped to deal with Iraq's
chemical attacks. Their stocks of protective gear were meagre, and
many of them would not even take the elementary precaution of
shaving their beards before wearing gas masks. Their strongest
weapon against Iraqi chemical warfare was essentially political—
namely, the propaganda value derived from the Iraqi attacks. Yet,
even in this respect they did not fare too well. At that time
Saddam was the favourite son of the West (and to a lesser extent
the Soviet Union), the perceived barrier to the spread of Islamic
fundamentalism. Consequently, apart from occasional feeble
remonstrations (for example, after the Halabja attack), the Western
governments were consciously willing to turn a blind eye to Iraq's
chemical excesses.

[22] See, e.g., *Daily Telegraph* (London), 4 Mar. 1988; *Financial Times* (London),
26 Feb. 1988; *The Times* (London), 22 Mar. 1988.
[23] Cordesman and Wagner, *Lessons of Modern War*, 517.
[24] See S. Carus, *Chemical Weapons in the Middle East* (Washington DC:
Washington Institute for Near East Policy, Dec. 1988), 4.

## THE MISSILE WAR

Saddam Hussein's initial war strategy, it may be recalled, was a strategy of limited war which sought to confine the war to the battlefield. For its part, Iran, recognizing its military inferiority on the ground, was quick to carry the war to the Iraqi rear, by initiating air and naval attacks on Iraqi strategic targets within twenty-four hours of the outbreak of hostilities. It was a defensive escalation aimed at avoiding defeat rather than achieving victory.[25] And it certainly achieved its objective as Saddam was forced to deviate from his original plans and to be drawn into counter-value exchanges which he had not initially intended.

The first Iraqi missiles were fired at the Iranian cities of Dezful and Ahvaz in November 1980. Those were the dated and inaccurate Frog-7s and the damage they caused was minimal. This did not deter the Iraqis from using these missiles extensively in 1981 (fifty-four missiles as compared with ten in 1980), but the employment of Frog-7s dropped significantly after the completion of the Iraqi withdrawal from Iran in mid-1982. Not only did Baghdad find growing difficulties in finding adequate targets for the Frogs, but its holdings of Scuds had by now reached the level that enabled it to use them on a regular basis. In 1983 Iraq fired thirty-three Scud missiles at Iranian cities, and this number grew to eighty-two in 1985.[26]

By this time, missile attacks had become an integral component of the overall Iraqi effort in the so-called Wars of the Cities, those ferocious exchanges against Iranian and Iraqi strategic targets and population centres. The role of missiles in these exchanges, however, was secondary, as aerial strategic bombardments took the lion's share of the Iraqi effort.

In the First War of the Cities (7–18 February 1984), merely a dozen Iraqi missiles were fired, and, although this number tripled in the Second War of the Cities (dragging on intermittently between March and July 1985), it plunged again in the Third and Fourth Wars of the Cities, which took place in 1987. It was only during the Fifth, and last, War of the Cities in 1988 that surface-to-

[25] See P. A. G. Sabin and E. Karsh, 'Escalation in the Iran–Iraq War', *Survival* (May–June 1989), 244–5.
[26] Unless stated otherwise, figures of missile attacks are taken from Cordesman and Wagner, *Lessons of Modern War*.

surface missiles came to play a key role in Iraq's strategy, though by no means exceeding the number of strategic bombings. At this stage, Iraq's stockpiles of ballistic missiles had become extensive, and it was making operational use of its home-made al-Hussein missiles. During March and April 1988 a barrage of some two hundred missiles rocked several Iranian cities, mainly Tehran. Skilfully combined with ferocious air attacks, the missile strikes were highly effective, leading to a mass exodus of Tehran residents and ultimately to the collapse of Iranian morale. This, in turn, paved the way for the Iraqi military successes of 1988, and to the Iranian agreement to stop fighting.

Though employed on a significantly smaller scale, missiles figured more prominently in Iran's war strategy than in that of Iraq. With the Iranian air force reduced within a year of the war's start to some eighty working aircraft, it was only natural that Tehran would come to consider ballistic missiles as its main strategic arm. Missiles were cheaper to buy and far more accessible than aircraft, and their maintenance and operation were significantly less complicated. Consequently, the 1985 War of the Cities witnessed the first significant use of Iranian ballistic missiles: fourteen missiles were fired against Iraqi cities, with at least three missiles landing in Baghdad. A similar amount of missiles was launched during the two Wars of the Cities of 1987.[27] As in the Iraqi case, Iranian attacks peaked during the Fifth War of the Cities, when no fewer than seventy-six missiles were fired at Iraq, sixty-one of them at Baghdad.[28]

However ferocious, the Wars of the Cities were conducted within broadly defined 'rules of the game', under which the two parties refrained from carrying their counter-value attacks to their logical extreme and tended to back down when confronted with resolute strategic retaliation. Consequently, major escalation did not take place until the Iranian gains of 1986.

Iranian caution reflected the formidable material and personal problems confronting the revolutionary regime. Baghdad's restraint before 1986 seems to have resulted from a mixture of considerations. It reflected a reluctance to antagonize Iraq's supporters, particularly Saudi Arabia and Kuwait, which feared that Iran

[27] MacNaugher, 'Ballistic Missiles', 9.
[28] The rest of the missiles were fired at Mosul (9), Kirkuk (5), and one at Tikrit, Saddam Hussein's home town.

might retaliate by widening the war. Saddam was also unwilling to jeopardize his air force, for serious losses (of pilots even more than aircraft) could have destroyed Iraq's most effective reserve force. Moreover, the ferocity of Iran's reaction to attacks on its population centres served as a powerful deterrent. Above all, Iraqi restraint was designed to prevent irreversible damage to future Iraqi–Iranian relations. Since Saddam fought the war with eyes set on post-war survival, he knew that boxing Iran into a corner could only complicate things for him in the future.

## NON-CONVENTIONAL WARFARE—HOW EFFECTIVE?

Since neither missiles nor chemical agents were employed in isolation from other weapons systems but were rather part of a wider and more comprehensive war effort, it is difficult to assess their direct operational utility. From the beginning of the war Iraq enjoyed a marked material edge over Iran, and this developed into overwhelming superiority with the passage of time, as the internationally isolated Iran barely succeeded in maintaining its pre-war arms holdings, whereas Iraq, bolstered by virtually the entire international community, managed to increase and improve its order of battle decisively. By the end of the war Iraq had come to enjoy a 10 to 1 superiority in most major weapons systems, including aircraft, tanks, and armoured vehicles. Iraq's mastery of the air was complete.

This, in turn, meant that chemical warfare was not that crucial for Iraq's war operations. Sheltered behind a formidable line of defence, the Iraqi troops could readily repulse the swarms of ill-equipped and poorly trained Iranians by relying on their overwhelming conventional superiority. Indeed, to judge by the fact that merely 2 per cent of Iran's total war casualties were caused by chemical weapons—some 27,500 injured and 260 dead out of nearly a million casualties[29]—the extent of Iraq's chemical warfare seems to have been more limited than is commonly assumed. As shown by the Halabja massacre, when used extensively against unprotected population, chemical weapons can have a devastating effect. Since most Iranian troops were no better

[29] These figures were released by the Iranians themselves in April 1988. See Carus, *Chemical Weapons in the Middle East*, 7.

equipped and prepared to contend with the consequences of a chemical attack than the hapless Kurds, a far higher rate of casualties than those reported by the Iranians (who, for their part, had a vested interest in inflating these figures) could have been expected, even while allowing for a measure of Iraqi incompetence (at least initially) in the use of chemical weapons.

Nor did gas prove the deciding factor in Iraq's offensive operations. Rather, the Iraqi success in turning the tables on Iran in 1987 and 1988 resulted from the growing professionalization of its armed forces following the loosening of Saddam's control over the conduct of the war. Free, by and large, of the political constraints imposed by their supreme leader, the Iraqi military managed to improve their fighting and to display an impressive level of combined-arms operations and manœuvrability, which allowed them to outnumber and outgun the Iranians in all military encounters. Take, for example, the liberation of the Fao Peninsula and the Majnoun Islands in the spring of 1988, two of Iraq's most shining military successes during that year. Both operations were carefully planned and competently executed, with the Iraqis enjoying an overwhelming numerical and material superiority over the demoralized and depleted Iranian forces, who showed little resistance to the advancing Iraqi forces. Consequently, 'chemical weapons probably added marginally to the impact of Iraq's massive tank force in the attack'.[30]

If Iraq's relatively restrained use of chemical agents during the defensive phase of the war was predominantly motivated by Saddam's anxiety not to push the Iranians beyond the point of no return, its sustained circumspection during the offensive phase of the conflict stemmed almost completely from pure military considerations. Chemical agents proved a delicate weapon to handle, and at times even a double-edged sword. There were reports of gas being blown back to the Iraqi positions due to miscalculation of weather conditions, as well as of shells and bombs that failed to explode being 'recycled' by the Iranians. Also, the less-than-perfect Iraqi intelligence and targeting capabilities advised against massive use of gas in close and complex encounters, of the sort that characterized the offensive phase of the war.

Interestingly enough, for most of the war the psychological

---

[30] MacNaugher, 'Ballistic Missiles', 18.

impact of chemical weapons was not greater than their pure military utility. As aptly noted by Tom MacNaugher, though there may be no objective reason 'why the use of high explosives for tearing men apart should be regarded as more humane than burning or asphyxiating them to death', most people would find the amorphous winds of death associated with the use of gas more grisly than other 'conventional' ways of dying.[31] And yet, it was not until the late stages of the conflict that the fearful spectre of chemical warfare came to dominate the battlefield. Even then, it had less to do with the intrinsic qualities of chemical weapons *per se* than with the collapse of Iranian national and military morale.

As long as the Iranians were largely united behind their political leadership and had the necessary zeal to prosecute the war, they were not too affected by Iraq's occasional resort to chemical warfare. It was only after the decisive majority of the Iranian public had come to recognize the futility of the unwinnable war that fears over the use of chemical weapons turned into widespread panic. This panic was further exacerbated by the Iranian authorities, who for their part sought to magnify Iraq's chemical threat in order to demonize the Iraqi enemy and to justify Iran's growing military setbacks. This propaganda ploy, nevertheless, backlashed. Not only did it fail to rally the masses behind the tottering war strategy of the regime, but it sent them fleeing Tehran and other Iranian cities for fear of Iraqi chemical attacks.

National morale was also the key factor in determining the efficacy of ballistic missiles. Until the 1988 War of the Cities, the impact of Iraqi missile strikes on Iranian cities was virtually nil. As noted earlier, missiles played a negligible role in the various Wars of the Cities in comparison to strategic bombing, and it was this latter form of warfare which gradually eroded Iranian national morale. Having achieved complete mastery of the air since the early stages of the war, Iraq could bomb Iranian strategic targets and population centres at will; and, even though the Iraqi air force proved to be a third-rate power, it nevertheless managed to deliver to its targets vastly larger payloads than Iraq's missile forces.[32] The decisive impact of missiles during the Fifth War of the Cities,

[31] Ibid. 23. His criticism is of M. van Creveld's assertion in *Technology and War: From 2000 BC to the Present* (New York: Free Press, 1989), 72, 177.
[32] See E. Karsh, 'Military Lessons of the Iran–Iraq War', *Orbis* (spring 1989), 216–17.

therefore, was the aggregate outcome of several unfortunate factors for the Iranians. With national morale at its lowest ebb, the fear of chemical strikes against population centres looming larger than ever, and material and economic dislocations increasingly unbearable, the combined air and missile attacks were the straw that broke the camel's back.

The impact of missiles on the Iraqi public was a mirror image of that of Iran. Unlike the Iranian regime, which, from the outset of the conflict, tried to unite the masses behind its campaign by stressing the virtue of sacrifice, Saddam sought to insulate the Iraqi population at large from the effects of the war. Fully aware of the fragility of Iraqi morale (after all, he had always had doubts as to how popular he really was among his own subjects), Saddam tried to prove to his people that he could wage war and maintain a business-as-usual atmosphere at the same time. The outcome of this guns-and-butter policy was that the ferocious war which raged on the battlefield was hardly felt on the Iraqi home front. Instead, the country was buzzing with economic activity and the population enjoyed an artificially high standard of living. Whenever the Iranians managed to puncture a hole in this artificial surrealist bubble by striking at Iraq's main population centres, as during the 1985 War of the Cities, Iraqi morale took an immediate plunge, driving an anxious Saddam to seek an immediate accommodation with the Iranians. Conversely, when the Iraqi public could eventually see the light at the end of the tunnel, its resilience to strategic attacks increased significantly. Thus, the seventy-six Iranian missiles that pounded Iraqi cities during the last War of the Cities had a far smaller impact on Iraqi morale than the sixteen missiles fired during the 1985 strategic exchange. This, in turn, underscores MacNaugher's conclusion that 'missiles are an effective *coup de grâce* weapon, decisive only when target states are already near collapse'.[33]

## CONCLUSION

Whatever their immediate operational value, chemical weapons and ballistic missiles emerged from the Iran–Iraq War as instruments of terror *par excellence*. The modest impact of these

[33] MacNaugher, 'Ballistic Missiles', 15.

weapons throughout most stages of the war has been conveniently overlooked by both sides, with the ominous visions of 1988 setting the standard for future reference. 'Chemical and biological weapons are the most important and most essential weapons of the world,' stated the Iranian President, Ali akbar Hashemi-Rafsanjani, shortly after the war. 'They are the poor man's atomic bomb and can easily be produced. We should at least consider them for our defence. Although the use of such weapons is inhuman, the war taught us that international laws are only scraps of paper.'[34] That Saddam Hussein holds a similar view is hardly at question. He has more than demonstrated his appreciation for non-conventional means by the famous threat to 'burn half of Israel' in April 1990, as well as by his televised interview with Peter Arnett during the 1991 Gulf War, in which he threatened to resort to chemical warfare should the need arise.

Given this regional state of mind, it is doubtful whether the proliferation of ballistic missiles and chemical weapons can be contained without an unwavering international resolve to do so. Even worse, while speaking of the 'poor man's atomic bomb', it is no secret that both Iraq and Iran are well embarked on an effort to achieve the 'actual thing'. And it is nuclear proliferation, particularly if successfully married to ballistic missiles, which, in the final account, poses the main threat to Middle Eastern stability, indeed to its very existence.

[34] Foreign Broadcast Information Service, *FBIS-NES*, 26 Sept. 1988, p. 26; 19 Oct. 1988, pp. 55–6.

# 3

# Non-Conventional Weaponry and Ballistic Missiles during the 1991 Gulf War

## MARTIN NAVIAS

SINCE the end of the 1991 Gulf War it has often been argued that neither ballistic missiles nor non-conventional weaponry had any effect on the outcome of Operation Desert Storm. Ballistic missiles proved militarily insignificant, chemical weapons were not employed, and Iraq's nuclear programme, though far more advanced and extensive than previously imagined, was not at the stage where it could produce an employable nuclear bomb. In the final analysis, what determined the result of the conflict was the qualitative conventional superiority of allied forces. That the result came so quickly and relatively painlessly was a reflection of the professionalism of allied troops, the standard of their technology, and the general incompetence of the Iraqis.

Yet, to deny the import of missiles and non-conventional weapons on the war's outcome is not the same as stating that these weapons did not have some important specific effects on the character of the confrontation. Indeed, it can be argued that the very existence of non-conventional weaponry had an impact upon the causes of the conflict, the conduct of the conflict, the conditions upon which the conflict was terminated, and the perception of that conflict in the Middle East and elsewhere. The same can certainly be said of ballistic missiles, which, unlike chemical weapons, were actually employed. It is argued here that none of these weapons was entirely inconsequential, even though they could not save Saddam from his folly. Equally as significant, these weapons may not necessarily be viewed by all as having played an inconsequential or irrelevant role, or a role that could not be suited to their various purposes. This is important, because the perception of the role of missiles and non-conventional weaponry

in the war may have consequences for the continued proliferation
of these items in the Third World in general and the Middle East
in particular.

### NON-CONVENTIONAL WEAPONS AND BALLISTIC MISSILES: WAR OBJECTIVES, PLANNING, AND OPERATIONS

One of the reasons for the 1991 Gulf War was the burgeoning
of Iraq's non-conventional and ballistic-missile programmes.
Obviously this was not the major reason (and it would not have
been a sufficient reason), but it was a not insignificant one in what
was sometimes referred to in the months before the war as the
allies' 'hidden agenda'. This agenda, many claimed, sought the
emasculation of Iraqi military, particularly non-conventional,
power and the neutralization of Iraq's long-range delivery
capabilities. In fact, during the period between August 1990 and
January 1991 this so-called hidden agenda became an increasingly
open one, as allied leaders sought to explain their build-up in the
Gulf as connected with the dangers of continued Iraqi proliferation.

Before 2 August, Iraq's non-conventional and ballistic missile
programmes had already become an object of growing concern,
because of reports of their advanced state, because the Iraqis had
demonstrated a willingness to use chemical weapons during their
war against Iran and the Kurds, and because of Iraqi boasts and
threats, which in April 1990 had included a promise to burn half
of Israel if that country dared attack Iraq. Statements by a number
of US officials concerned with proliferation matters indicated
that Iraq had become the number-one nuclear- and chemical-
proliferation concern in the developing world. In the wake of the
invasion of Kuwait this was a theme re-emphasized by the most
senior US officials, including the President and the Secretary of
State. For example, on 30 October US Secretary of State James
Baker emphasized that Saddam Hussein 'marries his old-style
contempt for civilized rules with modern destructive methods:
chemical and biological weapons, ballistic missiles, and—if he
could—nuclear weapons'.[1] On 22 November Bush went so far as

---

[1] 'Baker will not Rule Out Use of Force against Iraq', Speech at World Affairs
Council Meeting, 30 Oct. 1990, US Information Service Press Release.

to inform US forces in Saudi Arabia that, 'Those who would measure the timetable for Saddam's atomic program in years may be seriously underestimating the reality of the situation and the gravity of the threat.'[2] The destruction of Saddam's non-conventional (especially nuclear) capability and its attendant delivery systems became an increasingly important explanation for the war as mid-January approached. 'Fight now or pay later' was the motto of this rationale.

Many accused the President, in the words of two analysts writing later in the *Bulletin of Atomic Scientists*, of 'hyping the bomb'[3] in response to polls that showed that the destruction of Iraq's weapons of mass destruction was the one reason for which a majority of Americans believed a decision to go to war was justified. What, it was argued, the President was doing was legitimizing an action that was being undertaken with other objectives in mind. In order to do so, he was making out that the nuclear threat was worse than it actually was. Fear of Iraq's nuclear plans was certainly not unimportant in solidifying US public opinion behind the anti-Iraq alliance and thereby providing the President with more support and more room for manoeuvre than may otherwise have been the situation. Yet, whether Bush's 22 November statement was a hype, or a response at that stage to real information indicating that the Iraqi programme was more advanced than previously thought, is as yet not fully clear. Some reports have suggested that by the end of September 1990 the US Defence Intelligence Agency had received new intelligence information from satellite photographs as well as other sources indicating that Iraq was six months away from the bomb.[4] Certainly, post-war revelations indicate that the Iraqis were more advanced than previously thought.

This latter debate is, of course, separate from the question of whether the whole issue of Iraq's nuclear programme was being used to legitimize deployments in the Gulf. To agree with this argument would, however, be to ignore the enormous amount

[2] D. Albright and M. Hibbs, 'Hyping the Iraqi Bomb', *Bulletin of Atomic Scientists* (Mar. 1991).

[3] Ibid.

[4] K. R. Timmerman, *The Death Lobby: How the West Armed Iraq* (London: Fourth Estate Ltd., 1992), 390.

of attention and concern that Saddam's nuclear programme had caused in the West. Saddam's non-conventional-weapons programme had not led to allied deployments in the Gulf, but the crisis had underlined the future dangers of a Saddam armed to the teeth with non-conventional weaponry. If it was not the original reason for war, it became a reason for war.

Once the war was under way this declaratory emphasis on the destruction of the Iraqi non-conventional-weapons and ballistic-missile arsenal continued. For example, in the United Kingdom, on 27 January, Minister of Defence Tom King claimed UN authority for the destruction of Iraq's military capability. He stated that forcing Saddam's forces out of Kuwait without removing his capacity to mount a future threat would be a betrayal of the allied forces.[5] On 27 February Prime Minister Major told the Commons that Iraq must lose its weapons of mass destruction.[6] Both as an objective and as a rationale of the war, the stated goal of the destruction of Iraq's arsenal of mass destruction did not lose its appeal.

Missiles and non-conventional weaponry were also explicitly referred to in the ceasefire. Thus, on announcing the suspension of hostilities on 28 February, Bush was explicit: 'This suspension of offensive combat operations is contingent upon Iraq's not firing upon any coalition forces, and not launching Scud missiles against any other country. If Iraq violates these terms coalition forces will be free to resume military operations.'[7] UN Security Council Resolution 687, which formally ended the conflict, called for Iraq to remove and destroy not only all its weapons of mass destruction under the supervision of a specially created UN commission, but its ballistic missiles with a range greater than 150 kilometres as well.[8] Again in the summer of 1991, when Iraq proved recalcitrant about revealing the extent of its non-conventional-weapons stocks, there were threats and rumours of renewed US air strikes. On 29 May 1991 President Bush unveiled his proposal to ban weapons of mass destruction in the Middle East—a proposal which included a freeze on the buying, producing, and testing of

[5] *Independent*, 28 Jan. 1991.
[6] *The Times*, 1 Mar. 1991.
[7] Statement by President Bush, *Cable News Network*, 28 Feb. 1991.
[8] T. Deen, 'UN Votes for Iraqi Demilitarisation', *Jane's Defence Weekly*, 13 Apr. 1991.

ballistic missiles 'with a view to the ultimate elimination of such missiles' from the region.[9] Thus clearly, in both the pre- and the post-war phase, the destruction of non-conventional weaponry and ballistic missiles was an important element in allied objectives, rationalizations, and policies.[10]

In terms of planning for the operation, the existence of Iraq's chemical and biological capabilities affected preparations. President Bush was certainly greatly worried by the possibility of chemical war. 'All the plans', he stated, 'were predicated on [Saddam's] using chemical weapons because he'd done it before.'[11] Chemical-warfare training was thus stepped up in a more intense manner than had ever previously been the case. A biological immunization programme was undertaken, and medical facilities both in theatre and at bases overseas were readied for large numbers of chemical casualties.

The threat of chemical fighting also heavily influenced public perceptions of the conflict, and soldiers in gas suits became the central image of the crisis—at least until the beginning of fighting, when it was replaced by that of Patriots and Scuds. The possibility of chemical warfare was constantly discussed, as were its effects and the various responses. In the public eye at least, the potential of chemical warfare gave the conflict a certain apocalyptic atmosphere that may have helped mask the true discrepancy in power that was to be revealed once fighting got under way. For example, in the United Kingdom a medical specialist in chemical warfare, Alistair Hays of Leicester University, warned that chemical weapons would cause horrific casualties amongst allied troops, that the noddy suits would rapidly result in heat exhaustion, and that the antidote provided to the troops was 'more a morale-booster than a life saver'.[12]

The Iraqis certainly sought to play up these fears of non-conventional warfare. In April 1990, before the onset of the crisis, Saddam had already warned that he would burn half of Israel with chemical weapons if attacked by that country. With the crisis under way, Iraqi officials stressed that they would be ready to

---

[9] *International Herald Tribune*, 30 May 1991.
[10] The preceding section is derived from M. S. Navias, *Saddam's Scud War and Ballistic Missile Proliferation* (London: Centre for Defence Studies, 1991).
[11] H. Sidey, ''Twas A Famous Victory', *Time*, 6 Jan. 1992.
[12] *Guardian*, 24 Aug. 1991.

employ all forms of weapons in the defence of their country. For example, in December 1990 Parliament speaker Sadi Mehdi said Iraq would use all weapons, including chemical weapons, in order to defend itself.[13] In an interview with Peter Arnett before the land campaign, Saddam intimated that he possessed chemical warheads for his missiles and nuclear weapons as well. In the months leading up to the allied bombing campaign, the Iraqis also tested a number of their modified Scud missiles, thereby not only readying their forces but also underlining their ballistic-missile capabilities. In the final analysis, however, these threats were not sufficient to undermine Western public support for the war, though they did create fears as to the course the war would take, and did enter the debate over the necessity of war as opposed to continued sanctions.

Key areas of the public debate concerned whether Saddam actually possessed nuclear weapons (deemed unlikely); whether he possessed chemically armed Scud warheads (possible but uncertain); whether he would employ chemical weapons (uncertain but possible), and whether he would attack Israel (quite possible). It is quite likely that the internal debate which this debate reflected reached similar conclusions. At the same time, another debate ensued as to the nature of allied responses. Here it was generally agreed amongst analysts and pundits that the allies would not escalate to nuclear-weapons employment if their troops were gassed. There was, however, much uncertainty as to what the Israeli response would be.[14]

On the operational level, the annihilation of Iraq's weapons of mass destruction and ballistic missiles was a very important objective. Certainly, during the initial phase of the bombing campaign, fixed and mobile missile sites were targeted. During the

[13] *Sunday Times*, 23 Dec. 1990.

[14] In an *Independent–Newsnight* poll in August 1990, in answer to the question: if Iraq uses chemical weapons against British and American troops, should we use chemical weapons against Iraq?, 41% agreed, while 51% disagreed; 38% of Conservatives agreed, while 44% of Labour voters polled agreed. To the question: if Iraq uses chemical weapons against British and American troops, should we use nuclear weapons against Iraq?, 28% agreed, while 61% disagreed; 33% of Conservatives agreed, versus 25% of Labour voters polled (*Independent*, 24 Aug. 1990). At the beginning of February a US Gallup Poll found that 45% of those asked would favour use of nuclear weapons if it might save the lives of US troops. Three weeks earlier only half that number had favoured such an option (*Jerusalem Post*, 4 Feb. 1991).

following days, nuclear, chemical, and biological research centres were attacked. In both the missile and the non-conventional target cases, allied claims of destruction proved too optimistic. On 20 January US Press briefers maintained that Iraq's chemical facilities and its nuclear plants had been 'gravely damaged'.[15] While Saddam's non-conventional-weapons projects were undoubtedly affected, the scale of the projects were as yet not fully comprehended: some of the bases stretched out for many square kilometres, many were well protected, being dug deep underground, and not all the sites and projects were known to allied planners.

Of more immediate import were the modified Scud missiles. On the eve of the US air campaign, Deputy Secretary of State Lawrence Eagleburger told senior Israeli officials that the United States would destroy all of Iraq's launchers within two days[16] and, indeed, on 20 January Schwarzkopf told reporters that allied air forces had already neutralized all thirty of Iraq's fixed launchers and possibly sixteen out of twenty of the mobiles, though he admitted it was difficult to confirm.[17] What Schwarzkopf seemed to be intimating was that 92 per cent of Iraq's launchers had been neutralized by the third day of the war. One Saudi official apparently even went as far as to claim that the Iraqis no longer had the capability to hit any major Saudi population centres using missiles, except Hafr al Batin.[18] As a result of allied anti-Scud efforts, by 28 January the United States was claiming that thirty Scud fixed sites had been attacked twice and 'no capability exists at fixed sites to launch Scuds'.[19] On 30 January Schwarzkopf stated that 'we have managed to destroy all of their fixed sites, thirty fixed sites, we're quite confident of that. We have destroyed all their major missile production facilities and the pilots have reported to us more than fifty kills today . . .'[20] By the first week of February the Iraqis may have begun to attempt to repair some of their damaged fixed launch sites, but these were once more effec-

---

[15] IISS, *Strategic Survey 1991–2* (London: Brassey's, 1992), 69.
[16] E. Rosen, 'Army and Security', *Ma'ariv* (Hebrew), 29 Mar. 1991.
[17] Schwarzkopf interview, Riyadh, 20 Jan. 1991. Earlier, however, General Powell was reported as saying that '35 Scud launchers remain untouched' (*Sunday Times*, 20 Jan. 1991).
[18] Quoted in UK Military Press Briefing, Riyadh, 21 Jan. 1991.
[19] US Military Press Briefing, Riyadh, 28 Jan. 1991.
[20] US Military Press Briefing, Riyadh, 30 Jan. 1991.

tively retargeted by allied air forces. As a result it seems that no fixed sites were used to launch missiles throughout the war and they were used primarily as a base for mobiles.[21]

However, as the days passed after the initial allied air attacks, and Iraqi missile launchings continued—albeit at a reduced rate—confidence in fixed launcher destruction was not matched by confidence in the number of mobiles being destroyed. The problem was that, while the fixed launcher sites were undoubtedly being neutralized, claims about mobile launcher hits soon outstripped original estimates of Scud launchers. And it quickly became apparent—an important lesson of the war—that even a small number of mobile launchers could cause much nuisance and disruption. The Iraqis in fact continued to fire off their mobile Scuds throughout the war.[22]

As a consequence, further allied claims of mobile-launcher hits were usually couched in general terms, such as 'there has been a steady reduction in fixed and mobile launchers' (23 January) and 'during the last week there has been a marked decrease in Scud activity which we attribute to the effectiveness of the nightly air force Scud Buster mission' (1 February).[23] On 16 February one US briefer maintained that, while he did not know how many launchers were left, the 'Iraqis have not been able to volley fire for a long time'.[24]

Since the end of the war it has emerged that special forces were able to destroy a number of mobiles and may have succeeded in pushing others eastwards. How many were destroyed and whether the decrease in firing rates was a result of special force operation is, however, unclear.

The inability to destroy all the mobiles was partly explained by the operational difficulties of finding what Schwarzkopf likened to a needle in a haystack.[25] The ability of the Iraqis rapidly to fire their missiles and hide them in an enormous area meant that detection and destruction were no easy tasks. On 12 February an allied briefer also admitted that some of the claims of mobile

[21] US Military Press Briefing, Riyadh, 8 Feb. 1991.
[22] UK Military Press Briefing, Riyadh, 19 Feb. 1991.
[23] US Military Press Briefing, Riyadh, 1 Feb. 1991.
[24] US Military Press Briefing, Riyadh, 16 Feb. 1991.
[25] US Military Press Briefing, Riyadh, 18 Jan. 1991.

Scud hits—which by now exceeded the twenty to forty launchers originally believed to be in the Iraqi possession—may have been a result of the destruction of decoys.[26]

Another of the arguments that the military put forward for why they could not rapidly destroy all the mobile launchers was that there were far more of them than were originally expected. On 23 January, when the US spokesman was asked whether the allies had underestimated Iraqi launchers, his reply that they had not been underestimated was accompanied with the rider that 'intelligence estimates are an ongoing process'.[27] The US briefer stated that the allies knew how many launchers the Russians had sold the Iraqis but what was not known was how many they had lost during the Iran–Iraq War and how many they had managed to put together using parts of other launchers. Estimates amongst different intelligence services differed 'and the range is very broad, very broad indeed, and I wouldn't guess'. On 27 February Schwarzkopf was to take issue with the Press on reports that he had, on 20 January, said that the allies had destroyed as many as sixteen of Iraq's twenty mobiles. 'I think', said Schwarzkopf, 'that is a slight misquote. I think what I said at the time was we knew that they had more than 20 mobile missiles—we had a wide variety of estimates. I would tell you quite candidly that nobody knows exactly how many mobile missiles he had. I also said at the time we expected that we had damaged 16 of them.'[28] On 27 February Schwarzkopf explained the problem of why pilot estimates of missiles destroyed outnumbered the number of missiles originally believed to be in the Iraqi arsenal by stating that, 'we went into this with some intelligence estimates that I think I have since come to believe were either grossly inaccurate or our pilots are lying through their teeth and I choose to think the former rather than the latter'. He also said that, 'we went in with a very very low number of these mobile-erector-launchers we thought the enemy had. However at one point we had a report that they may have ten times as many.'[29] Whatever the case, few Scud launchers were destroyed by allied air attacks. According to one estimate the

[26] US Military Press Briefing, Riyadh, 12 Feb. 1991.
[27] UK Military Press Briefing, Riyadh, 22 Jan. 1991.
[28] US Military Press Briefing, Riyadh, 27 Feb. 1991.
[29] US Military Press Briefing, Riyadh, 27 Feb. 1991.

Iraqis may have 270 missiles left, though the figure may be as high as four hundred.[30]

At the end of January Schwarzkopf stated that the allies had flown 1,500 missions against the Scuds. The Israelis, who had so far remained out of the war, were far from content about the number of missions allied air forces were flying against Iraqi missile batteries located in the expanses of western Iraq. Tension quickly arose between the Israelis and the Americans, with Jerusalem attempting to get the Americans to step up their strikes against missile sites. These efforts may indeed have led Washington in some instances to increase the raids against the sites.[31] Not surprisingly, all this was accompanied by frictions within the allied camp between military and civilian authorities. According to one report, on two occasions Secretary of Defence Cheney complained that he did not believe that the air force was doing enough to increase the effort dedicated to the anti-Scud campaign. Conversely, Pentagon officials were angry that early in the air campaign 15 per cent of the central command's air assets were dedicated against what was regarded as tactically worthless missiles.[32] US sources indicated that in late January one-third of all allied air combat sorties were directed against Scuds.[33] As UK Air Chief Marshall Hines explained on 31 January, 'we have had to divert very considerable effort, and I think rightly so, for largely political reasons, on to locating and attacking Scuds'.[34] Some reports have even suggested that the anti-Scud campaign may have led to a week's delay in launching the ground assault,[35] though whether this was indeed so is not entirely clear.

The ability of the Iraqis continually to pepper Israel and allied forces with missiles imposed a number of costs upon the allies. Anti-Scud search-and-destroy operation tactics involved long hours of flying, a high level of training, exemplary co-ordination, as well as enormous technological resources. Not all developing countries possess the capability to mount such operations over an extended area and an extended period—especially if aircraft have

---

[30] M. Eisenstadt, *Iraq's Ballistic Missiles* (Washington DC: Washington Institute for Near East Policy, 1991).

[31] E. Rosen. 'Army and Security', *Ma'ariv*, 29 Mar. 1991.

[32] 'The Secret History of the War', *Newsweek*, 18 Mar. 1991.

[33] E. Bonsignore, 'The Scud War', *Military Technology*, 2 (1991).

[34] UK Military Press Briefing, Riyadh, 31 Jan. 1991.

[35] 'The Secret History of the War', *Newsweek*, 18 Mar. 1991.

also to be devoted to other pressing missions. In other words, missile-battery location will be a far more challenging task for developing countries than it was for the United States. For all these reasons, developing countries may come to see ballistic missiles as a poor man's weapon to be used to terrorize enemy populations and confound stronger forces. Certainly, one lesson of the war may well be that surface-to-surface missiles are important weapons for developing countries to acquire.

These were not the only benefits that accrued to the Iraqis. Specifically, the initial Scud launchings were the first indications that the Iraqi forces had not been completely annihilated by the early allied air strikes. They were also the one means whereby the Iraqi forces could, during the allied air assault, hit back at their foes while remaining on the strategic and tactical defensive. In addition, missiles were proof that the Iraqis possessed a means of penetrating Israel's formidable air defence and of causing distress amongst the civilian population of that country. They were the prime means by which Saddam attempted to penetrate the allied air space, and the most successful missile attack of the war—on a US army building in the Gulf—resulted in the most American casualties in one incident during the conflict. The need to destroy missiles before they could be launched consumed a significant portion of allied air efforts, which otherwise might have caused more damage to Saddam's forces—forces that still remain in place and support him in power. The fear that Saddam may have developed chemical warheads for his missiles added a further major element of unpredictability to the situation, as did the possibility that Israel may have found itself forced to retaliate against the Scud attacks, with all that that implied for the cohesion of the allied coalition.[36]

An important question then—though one that admittedly is not really answerable in any definitive manner—is what would the view have been in Israel before the attack, first if Saddam had had no missiles but only aircraft capable of delivering chemical weapons and, secondly, if he had had missiles but was known not to possess chemical weapons? Arguably, in the final analysis, it was the threat of chemical attack that caused the greater con-

---

[36] The preceding section is derived from M. Navias, *Saddam's Scud War and Ballistic Missile Proliferation* (London: Centre for Defence Studies, 1991).

sternation. The unique horror associated with chemical weapons, and the particular experience of the Holocaust, combined to lead Israelis to view a chemical assault with ever-increasing dread. What missiles did, however, was to make the possibility of a chemical strike against the Jewish heartland seem more likely. Trust and confidence in the Israeli Air Force had always been very high, but now missiles presented an apparently unchallengeable threat. Before the first attacks, the knowledge of missiles reinforced the fears, but it was the possibility of chemical attack that caused the most concern. Once the attacks began, the conventional-missile assaults severely disrupted public life, with fears being exacerbated by concerns that Saddam would soon escalate to chemical-missile strikes. Conversely, had Saddam possessed only aircraft, the Israeli public perception would probably have viewed with some confidence the ability of the Israeli Air Force to destroy them *en route*, but the particular horror of potential chemical attack would have made certain that the public would have remained most uneasy. Had Saddam possessed only missiles and been known to possess no chemical weapons, then what would have exercised the Israelis was their new vulnerability, though again, arguably, concern for this would have been less than fear of a chemical attack. Yet, in the final analysis, while fear of chemical attack was perhaps greater than the fear of missile attack, it is difficult to tease out the two threats. Both were very bound up with each other with missiles already in the pre-17 January phase underlining a new-found sense of vulnerability.[37]

### DETERRENCE AND THE NON-EMPLOYMENT OF NON-CONVENTIONAL WEAPONS

A key question is why Saddam never employed chemical weapons either against Israel or against US forces. A number of competing theories have been advanced. It has, for example, been argued that the poor state of Iraq's chemical ordnance made it very difficult for the Iraqis to deliver. There have been reports from the UN investigation team indicating that the warheads were unevenly filled, which would have destabilized the missiles and adversely affected their accuracy. It was also believed probable that the

---

[37] Ibid.

ordnance would have been destroyed on re-entry.[38] While Saddam's artillery ordnance was certainly more workable, clearly there were problems here too. The state of the Muthanna site, for example, indicated that the ordnance was in less than perfect condition. It has also been suggested—with specific regard to the Kuwaiti front—that the speed of the allied advance and the inability of the Iraqi command-and-control system to locate targets meant that the Iraqis had little chance to employ their non-conventional weapons. There has also emerged a hypothesis that Saddam gave his commanders the order to employ chemical weapons, but that they were deterred from doing so as a result of warnings by the allies that they would be held criminally responsible for such employment. The most intriguing argument, however, is that Saddam himself was deterred from using non-conventional weapons by allied and Israeli threats of response and escalation. According to Israel Air Force Commander General Avihu Ben-Nun in April 1991, 'The fact that he didn't launch chemical weapons against us was only because he feared our retaliatory response.'

Some reports indicate that the allies made it privately known to Saddam that they would respond with nuclear and chemical weapons should chemicals be used against them.[39] A review of some key declaratory pronouncements indicates that much effort and care was taken to warn Saddam of the grave dangers he faced should he escalate to the use of non-conventional weapons. A review of these statements, however, also reveals that both the allies and the Israelis never specified exactly what form their responses would take. No one ever said that they would definitely employ weapons of mass destruction and in some instances there were explicit statements that they would not do so.

The strongest US public statement indicating that US reaction would be severe should Saddam employ non-conventional weapons came in the letter Bush wrote to Saddam but which Tariq Azziz

[38] J. Tagliabue, 'UN Finds Toxic-Tipped Scuds', *International Herald Tribune*, 11 Nov. 1991.
[39] C. Bellamy, 'Allies "Put Iraqis off Chemical War"', *Independent*, 29 Nov. 1991. A liaison officer with the Swedish military hospital had said that he had seen a two-hundred-page booklet outlining the allied plan for a chemical counter-offensive. He stated that, 'It was not expected that use of chemical weapons would be necessary, but there was scope for it if it was' (ibid.).

refused to accept at his 9 January meeting with James Baker in
Geneva. In the letter Bush warned the Iraqi President that

the United States will not tolerate the use of chemical or biological
weapons, support of any kind of terrorist actions, or the destruction of
Kuwait's oilfields and installations. The American people would demand
the strongest possible response. You and your country will pay a terrible
price if you order unconscionable actions of this sort.[40]

Non-conventional weapons were not mentioned as a response, but
what was emphasized was that retribution would be swift and
heavy.

This was only one of a number of US statements to the same
effect. Soon after the crisis had begun, on 14 August, Secretary
of Defence Richard Cheney warned that, 'It should be clear to
Saddam Hussein that we have a wide range of military capabilities
that will let us respond with overwhelming force and extract a
very high price should he be foolish enough to use chemical
weapons on United States forces.' The Americans also made it
known to the Iraqis that weapons of mass destruction could be
employed against them. In January 1991 Cheney did this when he
stated:

Were Saddam Hussein foolish enough to use weapons of mass destruction,
the US response would be absolutely overwhelming and devastating ...
I assume [Saddam] knows that if he were to resort to chemical weapons,
that would be an escalation to weapons of mass destruction and that the
possibility would then exist, certainly with respect to the Israelis, for
example, that they might retaliate with unconventional weapons as well.

The Secretary was also no doubt hoping that the spectre of what
amounted to Israeli extended deterrence would help deter attacks
on US forces as well.

The Israelis, of course, were never that explicit, but they made it
very plain that an attack on them would be met with an extremely
hard response, especially if the attack was with non-conventional
weaponry. 'Iraq will be harmed in a most serious way ... whoever
will dare to attack us will be attacked seven times more,' said
Prime Minister Shamir in December 1990. There were numerous
statements of that type by Israeli policy-makers, but, like the
allies, the Israelis never indicated unambiguously whether they

---

[40] *International Herald Tribune*, 14 Jan. 1991.

would escalate to non-conventional-weapons employment. After all, Israeli policy is never to confirm or to deny the existence of its non-conventional arsenal.[41]

The Israelis did not make explicit statements that they would not escalate to non-conventional-weapons employment. On the other hand, there were a number of pronouncements by both US and UK policy-makers which served to cast doubt on that possibility. For example, in late January, Vice-President Quayle maintained, 'If Saddam Hussein uses chemical weapons, one option for us is to overwhelm him with conventional weapons and not nuclear weapons. But I am not going to rule out any option.' This equivocation was also reflected in a number of statements by Prime Minister John Major. On 7 January Major stated that he thought it unlikely that the Iraqis would employ chemical weapons and that Saddam did not as yet possess nuclear weapons. He also denied that the United Kingdom would employ nuclear weapons in response to an Iraqi chemical-weapons attack, stating that, 'We have awesome airpower quite apart from that.' Again on 8 January, at the headquarters of the 4th Armoured Brigade, when asked whether, should Saddam use chemical weapons, the United Kingdom would employ nuclear weapons, he replied: 'We have plenty of weapons short of that. We have no plans of the sort you envisage', and he repeated, 'Our men will be protected against chemical weapons. I hope he won't do that. I hope he'll actually have the sense to make a peaceful withdrawal, but we have plenty of weapons short of those you mention.'

Major's explicit references to UK nuclear non-employment was severely criticized. When Bush was asked on 5 February, whether the United States would employ weapons of mass destruction in response to Iraqi chemical weapons, his reply was, 'It's better never to say what you may be considering.'[42] Major seemed to have broken that rule, but so too did a 'senior officer attached to the 7th Armoured Brigade' who told the *Observer* in October 1990 that the United Kingdom would go nuclear if its forces were gassed.[43] This statement gave rise to much comment at the time,

---

[41] 'In my opinion, we have an excellent response, and that is to threaten Hussein with the same merchandise' (Israeli Science Minister Yuval Neeman, July 1990).

[42] M. Bundy, 'Nuclear Weapons and the Gulf', *Foreign Affairs* (fall, 1991), 84.

[43] *Observer*, 30 Sept. 1990.

but seems to have been purposefully disingenuous. While probably not reflecting UK policy, it was possibly designed to counter the earlier pronouncements which cast doubt on the likelihood on nuclear employment. In short, the object was to add to uncertainty in the Iraqi mind and reinforce deterrence.

Of course, it is far from clear whether Saddam was paying any attention to the nuances and contradictions contained in these statements. It is quite possible that, like so much other allied declaratory policy, they had little effect on Iraqi decision-making. However, Saddam must have had some feeling as to what the United States and what the Israelis would have done had he employed chemicals. It is quite probable that he believed that Israel would have escalated to a chemical and even a nuclear response. It is probable that he was quite clear in his mind about the strength of a potential Israeli reaction, and of Israel's willingness to escalate. His decision not to employ ballistic missiles armed with chemical weapons reflected his desire to bring Israel into the war on the cheap. He must have known that a chemical attack on Israel would definitely have brought it into the conflict, but how it would have come in must have deterred him. His decision to refrain from chemical-weapons employment was not based solely on the specific threats that Israel made during the crisis, but also on statements it had made over the past few years with regard to such attacks on its territory; no doubt he was also influenced by an appreciation of Israel's record of assertive responses and willingness to repay enemy aggression with much harder reactions.

With regard to the United States, Saddam probably paid some heed to Bush's statements of ambiguous intent, but more important was his particular reading of US society and its past record of war-fighting. Saddam seems to have doubted the US willingness to fight as well as its readiness to take casualties and to stay the course for the protracted war he promised. His view of the workings of US society was one of a weak government subjected to the vagaries of a spoilt and irresolute public. He probably doubted that the Americans would escalate to non-conventional weaponry, but what he may have feared was that, should war come and he employ chemical weapons, it could force the Americans to escalate the objectives to include himself. From Saddam's perspective this was just as bad—if not worse—than if the Israelis had employed non-conventional weaponry against the Iraqi populace. Even if he was so callous, he could not gain any advantages by refraining from employment of chemicals in one theatre (Kuwait) while employing them in another (Israel). To do so could have invited the worst of all scenarios: an escalation by the Israelis to non-conventional weapons and an escalation of allied objectives to include himself.

Finally, what do the experiences of the war tell us about the future of proliferation in the Middle East? One can argue that, because the weapons did not determine the outcome of the conflict in any way whatsoever, neither non-conventional weaponry nor ballistic missiles proved themselves to be attractive weapons for developing countries. Essentially they are not worth the effort. If anything, they attract undue attention. In the case of Iraq, they also served to legitimize allied actions and bolster Western public support of allied strategy. Developing countries would do better to invest in more sophisticated conventional weapons and more robust command and control facilities. However, I do not believe that this is the major lesson of the war, nor will it be perceived in that way by states in the Middle East and elsewhere.

For, in the final analysis, Saddam's failure in the Gulf War was not a result of the failure of this or that weapons system, but a result of a totally misguided grand strategy. Indeed, the major lesson of the conflict against which all other lessons pale in significance is that developing countries—even relatively rich and well-armed ones such as Iraq—would do well not to attempt to try and take on the United States when it is cloaked in the mantle of UN self-righteousness, backed up by most of the rest of the world, and busy trying to exorcize Vietnam ghosts.

No weapons system can alter that disparity in forces. This does not mean that there are not other scenarios in which chemical weapons could play an important role. Intra-Third World conflicts are perhaps more likely and more amenable to chemical-weapons threats, deterrence, and employment than wars between developing states and superpowers. Moreover, the war demonstrated that Western publics are quite sensitive to the threat of chemical warfare. In a situation where the choices are less politically and morally clear-cut than was the case in the Gulf, then chemical arsenals could deter external power projection. With regard to their military effectiveness, nothing new was confirmed or disproved by the war. Uncertainty as to their effects—one of their basic deterrent qualities—remains.

If one accepts the argument that Saddam was deterred from chemical-weapons employment against Israel because he feared Israeli nuclear retaliation, then nuclear weapons proved themselves to have deterrent value in intra-Third World conflicts. It is, then, reasonable to assume that they will also have great deterrent value in other Third World conflagrations and thus should be acquired. Nuclear weapons in the hands of developing states could also deter Western interventionism. One of the arguments of the conflict when it came to nuclear weapons was to fight now or pay later. This is not to argue that, had Saddam actually deployed nuclear weapons, the allies would not have fought; but it is to suggest that the decision to do so would have been a far heavier one to take. With regard to ballistic missiles, they demonstrated themselves to be great disruptive if not terror weapons. The effect of Saddam's Scuds on Israel's civilian population did not go unnoticed in the Middle East. Since most Middle Eastern states will continue to have great difficulty in penetrating Israeli air space, missiles continue to be attractive weapons. Israel is not, of course, the only potential target. Ghaddafi has, for example, stated that, had he deployed missiles capable of targeting New York, the Americans would never have bombed Tripoli in 1986.

While, as a result of the war, tightened supplier controls may make non-conventional weapons and ballistic missiles more difficult to acquire, a reading of the war does not support the conclusion that these systems are irrelevant to the security policies of states in the Middle East and elsewhere.

# PART TWO

# PROLIFERATION: PROGRESS AND PROSPECTS

# 4

# Chemical-Weapons Proliferation in the Middle East

## JULIAN PERRY ROBINSON

THIS chapter is about chemical-warfare weapons,[1] namely weapons which, by virtue of their toxic effects, are subject to the 1925 Geneva Protocol, the international treaty to which most of the world is now party and which prohibits resort to both chemical and biological warfare.[2] The chapter has two underlying themes. Its basic argument is drawn from the relationship between them.

One theme is the difference between the political and the military utility of chemical-warfare weapons. Populations can be terrorized with poison gas, just as they can with other types of military weapon. But, whereas poison gas appears to be exceptionally powerful in this regard, for the majority of purely military tasks conventional weapons will, on technical and operational grounds, be the weapons of choice. Poison gas is, therefore, peculiar in that its value as a military weapon seems to be far surpassed by its value as a political weapon.

This odd dichotomy is enhanced by another characteristic of

[1] The term 'chemical-warfare weapon' is used in this chapter in contradistinction to 'chemical weapon'. This is because, in the military lexicons of not a few countries, 'chemical weapon' is a term that covers not only the weapons of chemical warfare—which is to say those weapons that work through toxicity—but also such things as smoke and incendiary weapons. Another special term used in this chapter is 'poison-gas weapons', by which is meant the sub-set of chemical-warfare weapons that is restricted to antipersonnel chemical-warfare weapons but excludes those that are designed solely to harass rather than to cause casualties. So it means chemical-warfare weapons other than tear gases and herbicides.

[2] The Protocol prohibits the use in war of 'gaz asphyxiants, toxiques ou similaires, ainsi que de tous liquides, matières ou procédés analogues'. For a recent commentary, see R. J. McElroy, 'The Geneva Protocol of 1925', in M. Krepon and D. Caldwell (eds.), *The Politics of Arms Control Treaty Ratification* (New York: St Martin's Press, 1991), 125–66. All major countries of the Middle East and North Africa are parties, with the exceptions of Algeria and Chad.

chemical-warfare weapons: protection against poison gas is both
feasible and, for rich-country armed forces, available. Against
an enemy lacking gas masks and training in how to use them
properly, the military utility of chemical warfare could become
very substantially increased. But civilian populations cannot be
protected to anything like the same degree as military ones, save
perhaps in small, rich, and disciplined countries.[3]

The second theme of this chapter is that, just as the world
appears to be witnessing a proliferation of chemical-warfare
weapons, so also is it witnessing a deproliferation. Some countries
may indeed be moving into chemical-warfare armament; but other
countries, in a fashion set by the United Kingdom in 1956, have
been moving out. Both Russia and the United States are now
firmly committed to destroying their stocks, which most likely
account for at least 95 per cent of current world holdings of
chemical-warfare weapons.[4]

The argument of this chapter is that the dichotomy is a motor
of deproliferation. It can be stated in the following proposition.
Offensive chemical-warfare capability, by virtue of its terroristic
and therefore coercive potential, affords a way-station on the
track leading to nuclear weapons. There is evidence to suggest that
some such consideration has in fact been driving manufacture of
nerve gas in the Middle East,[5] meaning that the talk of linkage
between nuclear and chemical weapons, prominent in Arab state-

[3] These assessments of military and civil antichemical protection were developed
at an expert workshop in Spain on 'Antichemical Protection, its Potential and its
Relation to the Spread of Chemical Weapons and their Elimination' (see *The ASA
Newsletter*, 26 (9 Oct. 1991), 1, 18–20; *Chemical Weapons Convention Bulletin*,
13 (Sept. 1991), 17, 18). Participating in the workshop were defence scientists and
military specialists from sixteen countries, among them China, Egypt, Israel,
Jordan, and Pakistan.

[4] In a confidential exchange of information on 29 Dec. 1989, the US and Soviet
governments disclosed to each other particulars of their stocks of chemical-warfare
weapons. According to a subsequent report from the US General Accounting Office
(*Arms Control: US and International Efforts to Ban Chemical Weapons*, GAO/
NSIAD-91-317, 30 Sept. 1991), 'the United States stated that it had 29,000 agent
metric tons, and the USSR stated it had 40,000 agent metric tons'. For a detailed
but unofficial account of those holdings, see Parts I and II of J. P. Perry Robinson,
'World Chemical-Weapons Armament', *Chemical Weapons Convention Bulletin*, 2
(autumn 1988), 12–18, 4 (May 1989), 15–22.

[5] On Israeli work on chemical and biological weapons during the 1950s and
early 1960s as a stop-gap pending Israeli nuclear weapons, see S. M. Hersh, *The
Samson Option* (London: Faber & Faber, 1991), 63–4, 136. On the mid-1970s
publicity given by Cairo officials to the idea of an Egyptian nerve-gas programme

ments during and since the January 1989 Paris Conference,[6] has not been mere rhetoric. But, so the proposition continues, the attempts to assimilate the new nerve-gas weapons into military forces and doctrine have stimulated a slowly developing appreciation, just as they apparently did in the United Kingdom in the 1950s, that weapons of such strongly constrained utility are in fact largely valueless, despite the fear with which they may be regarded by potential enemies. So, if there are benefits to be gained from giving the weapons up, even quite small benefits might justify the costs of doing so; the costs in terms of damage to the national security, in particular, could hardly be considered great.

Public-domain data on the subject being so scarce, this is not an argument that can be definitively established (or refuted). It points up, nevertheless, an interesting parallel between the United Kingdom in mid-1956 and Iraq in mid-1991, both countries then just over a decade into their nerve-gas programmes. The decision by the UK government in July 1956 to curtail its programme was surely driven by some such cost-benefit assessment.[7] Would not a similar calculation explain the conspicuous readiness of Iraq to co-operate with the chemical work of the UN Special Commission charged with implementing terms of the 1991 Gulf War ceasefire resolutions rather than, as with other areas of Special Commission business, to impede it?[8] Alternative explanations for the Iraqi

having been necessitated by Israeli nuclear weapons, see W. Beecher, 'Egypt Deploying Nerve Gas Weapons', *Boston Globe*, 6 June 1976, p. 1. On the Syrian nerve-gas programme as counterweight to Israeli nuclear capabilities, see W. A. Terrill, 'The Chemical-Warfare Legacy of the Yemen War', *Comparative Strategy*, 10/2 (Apr.–June 1991), 109–19. On the Iraqi programme as conceived in 1974 as possible route to military parity with Israel, see Special Correspondent, 'Terror Arsenal the West Ignored', *Independent* (London), 12 Sept. 1990, p. 9.

[6] The French government has not yet published, as it said it would, the proceedings of the Conférence de Paris sur l'Interdiction des Armes Chimiques attended by 149 states, more than half of them represented at ministerial level, 7–11 Jan. 1989. One place where the text of the Final Declaration can be found is *Chemical Weapons Convention Bulletin*, 3 (Feb. 1989), 12. For contrasting accounts of the conference, see R. Dumas [Minister for Foreign Affairs of France], *NATO Review* (Apr. 1989), 1–3; V. Adams, 'Chemical Weapons Conference in Paris', *Bulletin* [London: Council for Arms Control], 44 (June 1989), 6–7.

[7] The decision still awaits its historian. But see the allusions to it in G. B. Carter, 'The Chemical and Biological Defence Establishment, Porton Down, 1916–1991', *RUSI Journal* (autumn 1991), 66–74.

[8] See 'Iraqi CBW Armament and the UN Special Commission', *Chemical Weapons Convention Bulletin*, 13 (Sept. 1991), 21–2.

behaviour can, of course, be envisaged, but Saddam Hussein certainly appears content to abandon his nerve gas.

This chapter is suggesting, then, that it is misleading to regard nerve gas as the 'poor man's atomic bomb', even though that is exactly how some states may at first have seen it, and even though ballistic missiles armed with nerve gas might very well, under conducive circumstances, have the political impact of an atomic bomb, at least until they are actually used. The real value of chemical-warfare armament, albeit limited, is mostly of a different kind, and to think of it primarily in nuclear-weapons terms is to risk misjudging it altogether, and therefore to risk failing to comprehend the actual threat to security inherent in the proliferation of chemical-warfare weapons. Misguided counter-proliferation policies may result.

The analysis which now follows is in three sections. The first two treat the constraints and incentives that have influenced poison-gas programmes of the past, drawing in part from the seventy-five-year history of Middle Eastern chemical-warfare armament. One describes utilities of chemical-warfare weapons, while the other seeks to identify the real roots of present concern about chemical proliferation. The third section addresses what is known of current programmes in the region.

THE UTILITY OF CHEMICAL-WARFARE WEAPONS

The historical record shows that poison-gas weapons have been used in less than a dozen—apparently no more than that—of the three hundred or so wars this century. The record is summarized in Table 4.1. Between the First World War (1914–18) and the Iran–Iraq (1980–8), the list contains only wars of intervention and colonial repression fought in the Third World. Conspicuously the list does not include either the Second World War or the Korean War or the Arab–Israeli wars; nor does it include the counter-insurgency efforts of the UK Royal Air Force in the days of Empire, despite recurrent consideration by the UK military at the prompting of certain gas enthusiasts.[9] Poison-gas warfare is, in short, a rare occurrence.

---

[9] D. E. Omissi, *Air Power and Colonial Control: The Royal Air Force, 1919–1939* (Manchester: Manchester University Press, 1990), 160.

TABLE 4.1. *Authenticated episodes of poison-gas warfare since the First World War*

| Period | User | Chemical-warfare weapons used |
| --- | --- | --- |
| 1919 | UK forces intervening in the Russian Civil War | Mustard-gas artillery, etc. |
| 1925 | Spanish forces in Morocco | Mustard-gas aircraft bombs |
| 1930 | Italian forces in Libya | Mustard-gas aircraft bombs |
| 1934 | Soviet forces intervening against Muslim insurgents in Sinkiang | Mustard-gas aircraft bombs |
| 1935–40 | Italian forces in Ethiopia | Mustard-gas aircraft spraytanks and bombs |
| 1937–45 | Japanese forces in China | Mustard-gas and lewisite aircraft bombs, etc. |
| 1963–67 | Egyptian forces intervening in the Yemeni Civil War | Phosgene and mustard-gas aircraft bombs |
| 1983–88 | Iraqi forces in the Iran–Iraq War | Mustard-gas and tabun aircraft bombs, etc. |
| 1987–88 | Iraqi forces in Iraqi Kurdistan | Mustard- and nerve-gas aircraft bombs, etc. |

*Source*: Sussex/Harvard Information Bank on Chemical- and Biological-Warfare Armament and Arms Limitation.

That rarity must presumably say something about the military utility of chemical-warfare weapons. Iraq's extensive use of poison gas in the Iran–Iraq War, compounded by the consequent expectations of chemical warfare during the 1991 Gulf War, has led many influential commentators to attribute all kinds of military value to poison gas. The historical picture in Table 4.1 cautions us to treat such asseverations sceptically. Much has now been published about the military benefits which Iraq gained from chemical-warfare weapons in its war with Iran,[10] but exceedingly

---

[10] See, e.g., M. T. May, 'Military Analysis of the Use of Chemical Weapons [in the Iran–Iraq War]', in G. M. Burck and C. C Flowerree (eds.), *International Handbook on Chemical Weapons Proliferation* (Westport, Conn.: Greenwood Press, 1991), 85–117; L. Waters, 'Chemical Weapons in the Iran/Iraq War', *Military Review*, 70/10 (Oct. 1990), 56–63; M. Bar, 'Strategic Lessons of

little of it cites patently reliable sources of information. Maybe more than appears is in fact soundly based, but, until the expert intelligence-based assessments are published, we have no way of knowing that; and on so seminal a topic, credulity can be no substitute for evidence.

It must also be noted that the episodes listed in Table 4.1 are but a small fraction of the total number of instances in which resort to poison-gas warfare has been reported or otherwise alleged. The better-known instances are listed in Table 4.2. Most of these reports, however, turn out to lack independent verification, and many appear, on closer examination, frankly incredible. One can sometimes see in them the hand at work, not just of misinformed reporters, but of deliberate disinformers. Why, for purposes of vilification, some people might choose to spread stories of poison-gas warfare, regardless of truth, is easy to understand: the subject is an emotive one, lending itself very well, therefore, to the turning of opinion against people, groups, or countries; or, no less, to sensational journalism. Table 4.2 may suggest that chemical warfare is becoming commonplace around the world, but actually the trend which it displays may tell us nothing at all about chemical warfare.

A class of weapon that is rarely used is one for which military demand is evidently slight, unless it be valued for a war-deterring propensity; and, not since a brief period some seventy years ago, well before the nuclear age, has anyone seriously suggested that poison gas might be such a deterrent.[11] 'Highly specialized' would be one way of describing the military utility of these weapons. It is easy to see why they have limitations. Psychological and cultural

Chemical War: Historical Approach', *IDF Journal*, 20 (summer 1990), 48–55; T. L. MacNaugher, 'Ballistic Missiles and Chemical Weapons: The Legacy of the Iran–Iraq War', *International Security*, 15/2 (fall 1990), 5–34; [V. S. Forrest and Y. Bodansky], 'Chemical Weapons in the Third World, ii. Iraq's Expanding Chemical Arsenal', Task Force on Terrorism and Unconventional Warfare, House Republican Research Committee, US House of Representatives, 29 May 1990; Major J. Johnston, 'Chemical Warfare in the Gulf—Lessons for NATO?', *British Army Review*, 91 (Jan. 1990), 25–31; A. H. Cordesman and A. R. Wagner, *The Lessons of Modern War*, ii. *The Iran–Iraq War* (Boulder, Colo.: Westview Press, 1990), esp. pp. 513–18; W. A. Terrill, jr., 'Chemical Weapons in the Gulf War', *Strategic Review*, 14/2 (spring 1986), 51–7.

[11] For further discussion, see 'The Value of Chemical Weapons for Deterrence', an annex in J. P. Perry Robinson, 'Chemical Warfare Arms Control: A Framework for Considering Policy Alternatives', *SIPRI Chemical and Biological Warfare Studies*, 2 (1985), 109–16.

TABLE 4.2. *State chemical warfare since the First World War: Verified and unverified reports of the use of chemical-warfare weapons by state forces, 1919–1991*

| Period | Reported user | Occasion of reported governmental use of chemical warfare |
| --- | --- | --- |
| 1919 | United Kingdom | Intervention in Russian Civil War |
| 1920 | United Kingdom | Counter-insurgency in Iraq |
| 1925 | Spain | Colonial warfare in Morocco |
| 1925 | France | Colonial warfare in Morocco |
| 1930 | Italy | Colonial warfare in Libya |
| early 1930s | China | Internal fighting in Manchuria |
| 1934 | Soviet Union | Intervention in Sinkiang |
| 1935–40 | Italy | Colonial warfare in Ethiopia |
| 1936 | Spain | Civil war |
| 1937–42 | Japan | War with China: use in Manchuria |
| 1939 | Poland | Defence against German invasion |
| 1942 | Germany | War with the Soviet Union: use in the Crimea |
| 1945–49 | China | Civil war |
| 1947 | France | Counter-insurgency in Indo-China |
| 1948 | Israel | War of Independence: against Egyptian forces |
| 1949 | Greece | Civil war |
| 1951–52 | United States | Korean War |
| 1951–53 | Britain | Counter-insurgency in Malaya |
| 1957 | Cuba | Internal fighting: use by government forces |
| 1957 | France | Counter-insurgency in Algeria |
| 1958 | Spain | Counter-insurgency in Rio de Oro |
| 1958 | Taiwan | Firing into China from Quemoy |
| 1959 | United Kingdom | Intervention in Oman |
| 1961–67 | United States | Intervention in South Vietnam |
| 1963–67 | Egypt | Intervention in the Yemen |
| 1965 | Iraq | Internal fighting against Kurdish rebels |
| 1967–70 | North Vietnam | Intervention in South Vietnam |
| 1968 | Portugal | Counter-insurgency in Guinea-Bissau |
| 1969 | Israel | Internal fighting against Palestinian guerrillas |
| 1970–73 | Portugal | Counter-insurgency in Angola |
| 1970–79 | Rhodesia | Internal fighting against ZAPU/ZANU |
| 1972–73 | Portugal | Counter-insurgency in Mozambique |

TABLE 4.2. *Continued*

| Period | Reported user | Occasion of reported governmental use of chemical warfare |
|---|---|---|
| 1975–84 | Laos | Internal fighting: use of 'yellow rain', etc. |
| 1976 | Morocco | Internal fighting against POLISARIO |
| 1977 | Zaïre | Internal fighting: use of poison arrows |
| 1978–87 | Vietnam | Intervention in Cambodia: use of 'yellow rain', etc. |
| 1978 | South Africa | Raid on SWAPO base in Angola |
| 1979 | Vietnam | Defence against Chinese invasion |
| 1979 | China | Invasion of Vietnam |
| 1979–88 | Soviet Union | Intervention in Afghanistan: use of 'yellow rain', etc. |
| 1980 | Ethiopia | War with Somalia |
| 1980–82 | Ethiopia | Internal fighting against Eritrean secessionists |
| 1980–88 | Iraq | War with Iran |
| 1981 | South Africa | Raid on SWAPO base in Angola |
| 1981–85 | El Salvador | Internal fighting |
| 1982 | Israel | Intervention in Lebanon |
| 1982–86 | Thailand | Firing into Cambodia |
| 1982–85 | Burma | Internal fighting |
| 1983–86 | United States | Covert intervention in Nicaragua |
| 1983 | United States | Intervention in Grenada |
| 1984 | Philippines | Internal fighting: use in Mindanao |
| 1984–86 | South Africa | Intervention in Namibia |
| 1985 | Indonesia | Internal fighting against FRETILIN in East Timor |
| 1985–86 | Nicaragua | Internal fighting: use of 'yellow rain', etc. |
| 1985–91 | Angola | Civil war: use against UNITA |
| 1986 | Mozambique | Civil war: use against RENAMO |
| 1986 | Chad | Civil war |
| 1986–87 | Libya | Intervention in Chad |
| 1986–88 | Iran | War with Iraq |
| 1986–89 | Afghanistan | Internal fighting |
| 1987–88 | Iraq | Suppression of Kurdish revolt |
| 1987–88 | South Africa | Intervention in Angola |
| 1988 | Thailand | Firing into Laos |
| 1988 | Libya | Intervention in Sudan |
| 1988 | Turkey | Internal fighting against Kurds |
| 1988–89 | Cuba | Intervention in Angola |
| 1988–90 | Israel | Internal fighting: suppression of the *Intifada* |

TABLE 4.2. *Continued*

| Period | Reported user | Occasion of reported governmental use of chemical warfare |
|--------|---------------|----------------------------------------------------------|
| 1989 | Romania | Internal fighting |
| 1989 | Somalia | Internal fighting |
| 1989–90 | Sudan | Internal fighting |
| 1991 | Iraq | Suppression of Shi'ite and Kurdish revolts |
| 1991 | India | Internal fighting in Kashmir |
| 1991 | Yugoslavia | Internal fighting |

*Source*: Sussex/Harvard Information Bank on Chemical- and Biological-Warfare Armament and Arms Limitation.

factors engendered by their unique mode of action—poisoning—translate into legal and political constraints on use. Technical constraints reside in the peculiarity that most chemical-warfare weapons work, not through direct action on their targets (as bullets or explosives do), but indirectly, by polluting the environment of the target. Military constraints, too, stem from this feature, because its consequence is a poor predictability of outcome, inimical, therefore, to tight forward-planning and the concerting of force in the field. Also, the peculiar mode of attack demands special skills, equipment, and training, the provision of which must inevitably impose weighty opportunity costs upon overall military capability.

It is the environmental mediation, furthermore, which means that antipersonnel chemical-warfare weapons are relatively easy to protect against: a filter interposed between the air a person has to breathe and his or her nose and mouth; overgarments to shield the skin from any rain of liquid chemical-warfare agent that might otherwise fall upon it. These are technologically demanding requirements if the wearers of the protection are to remain efficient in their work while protected, and are especially so in hot humid weather; but not nearly as demanding as comparable protection against blast, heat, or high-energy fragments.

Chemical-warfare weapons, in short, have a singularly broad range of costs associated with their use. The opportunity costs, in particular, may always be judged unacceptably high where the potential targets of chemical attack have antichemical protection of the kind just described. This, indeed, seems to be the principal lesson of Table 4.1. In all the instances noted there, the chemical warfare was initiated against an unprotected adversary. Such technology dependence of the opportunity costs also extends to the technology available to the potential user. For example, the greater his assimilation of modern techniques and equipments for environmental sensing, communication, target-acquisition, and the like, the smaller will be the additional burden imposed upon his forces and their infrastructure by preparedness to use chemical-warfare weapons. One may broadly conclude, then, that the limitations of chemical-warfare weapons will be at their least constraining for armies of the North and for those of the South that are adept at attracting and absorbing technology transfers. For less well-endowed forces, the incentives to use the weapons would surely have to be very great indeed for the opportunity costs, to say nothing of political and financial costs, to seem a price worth paying.

What might those incentives be? From study of the episodes noted in Table 4.1, the following conclusion seems tenable. Where poison-gas weapons are militarily attractive, their attractions reside in a 'force-multiplying' effect and in a morale effect. Their effects on morale need no explanation: the record is clear on the psychological impact of poison and the terror it can instil into military as well as civilian populations—an impact which stems, no doubt, from the same roots as does the taboo against using the weapons. The 'force-multiplier' effect depends for its nature and magnitude on the antichemical protection ranged against the weapons. If there is no protection, the multiplication results from the economy of force available from an area weapon, especially one that can reach inside fortifications, as a cloud of vapour or aerosol may be able to do, or remain active for extended periods of time, as liquids of low vapour pressure can after being sprayed over surfaces. If there is protection, the multiplication is the much less dependable one that stems from whatever degradation of the adversary's combat performance may result from having to fight within encumbering protective gear or being forced to avoid areas that cannot be decontaminated.

How great the protection-driven degradation of fighting efficiency might be is a matter on which there is as yet little recorded recent historical experience: only exercise- and manœuvre-derived information.[12] Desert Storm forces, however, were in full anti-chemical protection when the ground offensive was launched on 24 February 1991; the assessments of what impact that protection had remain, at the time of writing, to be published.

In 1947, a special commission of the United Nations categorized chemical-warfare weapons, for organizational reasons, alongside those of atomic, biological, and radiological warfare, as 'weapons of mass destruction'. The designation has persisted, taking on a life of its own, and is perhaps why so many people have seen nothing strange in poison gas being described as the 'poor man's atomic bomb'. Yet to what degree might chemical-warfare weapons actually display 'mass destructiveness'? Some chemical-warfare weapons—for example, tear gases and other incapacitants—mostly have transient effects if used in prescribed densities, in which case they are not, on their own, weapons of any sort of destructiveness. Chemical-warfare weapons with permanent effects do not destroy inanimate objects, but among people they can produce casualties on a massive scale; and have done so.

The nerve gases, in particular, are of such a deadliness that about five tons, which is perhaps a hundredth of a per cent of what today's arsenals contain, could, according to a reliable estimate, kill just about everyone in the world, if everyone lined up for injection. The corresponding figure for hydrogen cyanide could be about 250 tons; for bullets, 50,000 tons. How deadly nerve gas would be in practice is a matter of conjecture, though the fate of the Kurdish citizens of Halabja in March 1988, apparently attacked with sarin, gives indication. It would depend on what assumptions are made about means of delivery, about the prevailing weather conditions, and about the state of protectedness of the target. Those five tons of nerve gas, loaded into suitable munitions, would take maybe two strike aircraft to deliver—or thirty-six of the chemical warheads which Iraq had for its extended-range Scud missile, the Al-Hussein. According to one of the more widely used atmospheric-dispersion models, five tons of nerve gas, say sarin, thus disseminated over open country,

[12] On which, see especially M. S. Meselson, keynote address at the 6th Annual Scientific Conference on Chemical Defense Research, US Army Chemical Research, Development and Engineering Center, 13–16 Nov. 1990. The address is excerpted in *Chemical Weapons Convention Bulletin*, 11 (Mar. 1991), 17–19.

would threaten 50 per cent casualties across some four square kilometres, give or take a factor of three or four depending on the weather, provided the inhabitants of the area had no special protection; in the different meteorology of an urban environment, the equivalent area would be smaller, maybe only half as great.[13] There would be casualties further downwind, but in diminishing numbers and of diminishing severity. From data gathered after the atomic-bomb attack on Hiroshima, it is estimated that, for a relatively small airburst fission bomb (10–15 kt yield), the 50-per-cent casualty contour would encompass some thirty square kilometres.

In other words, poison gas could be comparable in destructiveness to the smaller sorts of nuclear weapon. Or, under different circumstances of weather or target posture, it might cause no great damage at all. The industrial chemical accident at Bhopal in December 1984, which killed more than two thousand people outright and injured hundreds of thousands more, involved the release of an airborne poison in a quantity much the same (in terms of numbers of theoretical lethal doses) as that in a comparable accident in Hamburg in May 1928, during which, thanks to the weather, no more than eleven people died. The quantity of sarin nerve gas equivalent to the amount of poison released in either case could have been carried in a single warhead for an unextended-range Scud missile. It is the contrasting certainty of effect of nuclear weapons—that, and the altogether greater magnitude of effect—which places them in so different a category.

All radically new forms of weaponry have had to survive an initial period of military disfavour and moral opprobrium before becoming assimilated into forces and doctrine—before their full range of military utilities becomes appreciated for what it is. Some novel weapons never make it. It is clear from their history that chemical-warfare weapons, despite their antiquity, are still located

---

[13] These figures are from J. P. Perry Robinson, 'Quantitative Estimates of the Possible Effects of CBW Attacks on Civilian Targets', October 1969, which is the unpublished 'Appendix 3' referred to in SIPRI, *The Problem of Chemical and Biological Warfare*, ii. *CB Weapons Today* (Stockholm: Almqvist & Wiksell, 1973), 143, 159, 271, 273. This study elaborated the work of an expert group initially convened by the World Health Organization for its *Health Aspects of Chemical and Biological Weapons* (Geneva: WHO, 1970). More recent estimates, in fact very similar ones, can be found in S. Fetter, 'Ballistic Missiles and Weapons of Mass Destruction', *International Security*, 16/1 (summer 1991), 5–42.

within this initial phase. They are held back from assimilation by a deep-rooted cross-cultural taboo on poison weapons, and by the limitations on military utility imposed primarily by the bio-specificity of their action, but also by other technical character-istics. It can be seen, too, that their assimilation has been pushed back by antichemical protective technologies and also by emergent weapons technologies which offer target effects previously unique to poison gas; but impelled forwards, nevertheless, by a peculiar combination of apparent cheapness and terroristic potential.

### THE ISSUE OF CHEMICAL PROLIFERATION

Those last two attributes are ones for which demand evidently exists in several conflict-torn parts of the world, creating at least some pressure for the weapons to spread. The taboo, operating through international law and custom, may weaken the pressure; but a world that has come to live with the doctrines and practices of nuclear deterrence is a world in which terrorization has been legitimized, taboos upturned. The attractions which chemical-warfare weapons hold out may, to the military, be diminishing, but to some political leaderships they may, in contrast, be in-creasing. It might well be supposed, then, that the whole issue of chemical proliferation is strongly coloured, maybe even its future determined, by that dichotomy between the political and the military utility of chemical-warfare weapons.

There is one other political dimension to chemical-warfare armament that should not be disregarded: chemical-warfare weapons have been available for use in wars more numerous than those in which they have actually been used. This suggests that levels of chemical-warfare armament have been driven more by supply than by demand. It is well known that the political necessity of making supply and demand appear to match one another can induce exaggerated perceptions of the military value of any armament that is in fact supply-led. There is no reason to suppose that chemical-warfare weapons are immune to this influence.[14] A prudent person will, therefore, ask whether the

---

[14] On the contrary. See, e.g., J. P. Perry Robinson, 'Supply, Demand and Assimilation in Chemical-Warfare Armament', in H. G. Brauch (ed.), *Military Technology, Armaments Dynamics and Disarmament* (London: Macmillan, 1989), 112–23.

present propagation of belief, remarked earlier—that chemical-warfare weapons are more useful than their history would otherwise suggest, that they are proliferating, and that their proliferation is real cause for concern—is not in some degree a manifestation of this influence. Institutions, even ones as unpopular as military chemical warfare services, can command surprising power once they become adept in the bureaucratic politics of national security.

If chemical-warfare weapons are thought of as the 'poor man's atomic bomb', their proliferation is obvious cause for concern. Spread of the weapons would pose problems for crisis stability, arms control, and international security above all in the Middle East, where the spread of weapons of mass destruction must surely exacerbate regional tensions and may, in time of crisis, accelerate the momentum towards full-scale conflict, deterrence notwithstanding.

Yet chemical-warfare weapons have been in the Middle East and associated regions of the Maghreb and contiguous parts of Africa for a long time. At the end of the Ottoman Empire, Turkey was believed by UK forces in Mesopotamia in 1916 to have deployed chemical-warfare weapons, in the form of poison-gas cylinders supplied by Germany, to Baghdad.[15] This (correct) belief seems to have been the initial reason for those forces being kept supplied, from 1917 on into the post-First World War period, with chemical-warfare weapons from the United Kingdom.[16] UK artillery units fired poison gas during the second Battle of Gaza in April 1917.[17] The chemical-warfare artillery munitions held at the UK garrison in Baghdad were used against Shia insurgents in the summer of 1920 'with excellent moral effect'.[18] In north Africa, a factory for mustard-gas weapons was built not long afterwards in Morocco by a private, subsequently notorious, German firm.[19]

[15] F. J. Moberly, *The Campaign in Mesopotamia, 1914–1918*, iii (London: HMSO, 1925), 14; C. H. Foulkes, *'Gas!' The Story of the Special Brigade* (Edinburgh: Blackwood, 1936), 95.

[16] F. J. Moberly, *The Campaign in Mesopotamia, 1914–1918*, iv (London: HMSO, 1927), 49; PRO WO 32/5184, WO 32/5191.

[17] G. MacMunn and C. Falls, *Military Operations, Egypt and Palestine* (London: HMSO, 1928), 336–7, 349.

[18] Omissi, *Air Power and Colonial Control, Observer*, 160; and D. Omissi, 'RAF Officer who Resigned rather than Bomb Iraq', (London), 10 Feb. 1991, p. 10.

[19] R. Kunz and R.-D. Müller, *Giftgas gegen Abd el Krim: Deutschland, Spanien und der Gaskrieg in Spanisch-Marokko, 1922–1927* (Freiburg im Breisgau: Verlag Rombach, 1990).

Later, in the 1930s, Italian air forces operating in Libya and then in Ethiopia had supplies of chemical-warfare weapons, some of which, as is well known, they used. There are state papers in Italian archives which cite the existence of UK chemical-warfare weapons in Kenya (though not elsewhere, apparently) as reason for the supplies.[20] Further proliferation occurred during the Second World War with the deployment of rather large stocks of chemical-warfare weapons into the region.[21] It is not at all clear from the public record that these stocks were all withdrawn once the war was over. The chemical-warfare weapons used in the Yemen during 1966–7, and perhaps earlier too, by Egyptian air forces intervening in the civil war are thought to have come from Second World War UK supplies, though other sources have also been mentioned.[22] The Yemeni republicans themselves are said to have received chemical-warfare weapons from China—phosgene aircraft bombs originating in Second World War lend-lease shipments, their US markings still discernible. So it is not a new phenomenon, the spread of chemical-warfare weapons to the Middle East. Nor is the special concern about chemical-warfare ballistic missiles. On 20 March 1963, for example, Israeli Prime Minister Golda Meir spoke in the Knesset of the dangers residing in Egyptian work on such weapons.

The Middle East is certainly part of the reason why chemical proliferation is now a prominent security issue. Yet it is clearly not the only reason. What else has contributed, and are they factors which themselves bear upon chemical-warfare armament in the region?

In fact, many different concerns have found expression. Voices warning about chemical proliferation have been heard since at least 1970—for example, in testimony taken by the US Congress[23]

[20] A. Sbacchi, 'Legacy of Bitterness: Poison Gas and Atrocities in the Italo-Ethiopian War, 1935–1935', *Genève-Afrique*, 13/2 (1974), 30–53.

[21] e.g., as at 1 Apr. 1945 UK stocks of agent-filled chemical-warfare munitions in the Middle East theatre comprised some 24,000 mustard-gas aircraft bombs (30 lb, 65 lb, and 250 lb), 500 mustard-gas spraytanks (250 lb and 500 lb), 290,000 mustard-gas artillery shells (25 pdr and 5.5 in.), 72,000 tear-gas artillery shells (25 pdr and 5.5 in.) and 4,000 phosgene mortar bombs (4.2 in.). See PRO CAB 128/773.

[22] For a review, see SIPRI, *The Problem of Chemical and Biological Warfare*, i. *The Rise of CB Weapons* (Stockholm: Almqvist & Wiksell, 1971), 161.

[23] M. S. Meselson, 'Policy Considerations Regarding the Use of Harassing Gas in War', prepared statement before the Defense Subcommittee, Committee on Appropriations, US Senate, 20 May 1970; J. P. Perry Robinson, prepared

and during workshops of the Pugwash chemical-warfare Study Group from 1974 onwards.[24] A doctoral dissertation on the subject was also completed at this time,[25] followed intermittently by analyses from other non-governmental quarters.[26] But it was not until April 1984, shortly after the UN Secretary General had verified use of poison gas in the Gulf War, that minatory *governmental* statements about chemical-warfare proliferation began to accumulate in the public record. The US Defense Intelligence Agency started it all with testimony to the Senate: 'most of the threat', Mr Dominic Gasbarri told the Armed Services Strategic and Theater Nuclear Forces subcommittee on 26 April 1984, 'has been with the Soviets, but we now have evidence that indicates other countries want chemical weapons.' The details, such as whether or how the wants might be being satisfied and who was thought to have them, furnished by his colleague Captain Sylvia Copeland, were deleted from the testimony as published. But investigative reporting of chemical-warfare proliferation commenced in the news media shortly afterwards,[27] stimulated by

statement before the Subcommittee on National Security Policy and Scientific Developments, Committee on Foreign Affairs, US House of Representatives, 2 May 1974, hearings, *US Chemical Warfare Policy*, 63–9.

[24] *Pugwash Newsletter*, 11/5 (June 1974), 115–72, 13/4 (Apr. 1976), 188–202, 15/3 (Jan. 1978), 84–95, 24/2 (Apr. 1987), 108–11, respectively the reports of the First (Helsinki), Third (London), Fifth (Cologne/Leverkusen), and Twelfth (East Berlin) workshops.

[25] J. S. Finan, 'Chemical and Biological Weapons: Their Potential for Nations outside the Principal Alliances, with Special Reference to the Possibilities Open to the Republic of South Africa over the Next Ten Years' (Ph.D. thesis, University of London, 1975).

[26] See, e.g., G. K. Vachon, 'Chemical Disarmament—A Regional Initiative?', *Millennium: Journal of International Studies*, 8/2 (1979), 145–54; J. P. Perry Robinson, 'Chemical, Biological and Radiological Warfare: Futures from the Past', submission to the Palme Commission, Sept. 1981; G. K. Vachon, 'Chemical Weapons and the Third World', *Survival* (London: IISS), 26/2 (Mar.–Apr. 1984), 79–86; B. Roberts, H. Lin, W. Donnelly, and J. F. Pilat, 'Binary Weapons: Implications of the US Chemical Stockpile Modernization Program for Chemical Weapons Proliferation', a report prepared by the Library of Congress Congressional Research Service for the House Foreign Affairs Subcommittee on International Security and Scientific Affairs, 24 Apr. 1984.

[27] Above all R. Halloran, 'US Finds 14 Nations Now have Chemical Arms', *New York Times*, 20 May 1984, p. 22; J. Anderson, 'The Growing Chemical Club', *Washington Post*, 26 Aug. 1984, C7; L. R. Ember, 'Pentagon Pressing Hard for Binary Chemical Arms Funds', *Chemical and Engineering News*, 25 Feb. 1985, pp. 26–8; D. Oberdorfer, 'Chemical Arms Curbs are Sought', *Washington Post*, 9 Sept. 1985, p. 1; L. R. Ember, 'Worldwide Spread of Chemical Arms Receiving

leaked official papers and unattributable official briefings, and both sustained by and sustaining a motley of academic and political commentators.[28] Issue creation was now well and truly under way.

Increased Attention', *Chemical and Engineering News*, 14 Apr. 1986, pp. 8–16; R. C. Toth, 'Germ, Chemical Arms Reported Proliferating', *Los Angeles Times*, 27 May 1986, p. 1; R. Harris and P. Woolwich, 'The Secrets of Samarra', BBC *Panorama* television documentary, screened in Britain, 27 Oct. 1986; J. Smolowe, 'Chemical Warfare: Return of the Silent Killer', *Time Magazine* [US edn.], 22 Aug. 1988, pp. 46–9; J. J. Fialka, 'Fighting Dirty', *Wall Street Journal*, 15 Sept. 1988, pp. 1, 24; 16 Sept. 1988, pp. 1, 22; 19 Sept. 1988, pp. 1, 30; R. Wright, 'Chemical-Arms Race Heating Up', *Los Angeles Times*, 9 Oct. 1988, pp. 1, 6, 8; G. Thatcher, with T. Aeppel, P. Grier, and G. D. Moffett III, 'Poison on the Wind, pt. 1. The Poisons Spread', *Christian Science Monitor*, 13 Dec. 1988, B1–B16; H. Koppe and E. R. Koch, *Bomben-Geschäfte: Tödliche Waffen Für die Dritte Welt* (Munich: Knesebeck & Schuler, 1990), 223–89; H. Leyendecker and R. Rickelmann, *Exporteure des Todes: Deutscher Rüstungsskandal in Nahost* (Göttingen: Steidl Verlag, 1990, 1991).

[28] See, esp., J. D. Douglass, jr, and N. C. Livingstone, 'CBW Proliferation: An Even More Dangerous World', in *America the Vulnerable: The Threat of Chemical and Biological Warfare* (Lexington, Mass: D. C. Heath & Co, 1987), ch. 5; B. Roberts (ed.), 'Chemical Warfare Policy: Beyond the Binary Production Decision', *Significant Issues Series* [Washington DC: Center for Strategic and International Studies, Georgetown University], 9/3 (May 1987), 36–40; J. P. Perry Robinson; 'Some Developments over the Past Year in the Field of Chemical-Warfare Armament', paper for the 13th Workshop of the Pugwash Chemical Warfare Study Group, Geneva, 22–4 Jan. 1988; W. S. Carus, 'Chemical Weapons in the Middle East', *Policy Focus* (Washington Institute for Near East Policy, Research Memorandum 9; Dec. 1988); E. M. Spiers, *Chemical Weaponry: A Continuing Challenge* (London: Macmillan, 1989), 131–44; E. D. Harris, 'Chemical Weapons Proliferation in the Developing World', *RUSI and Brassey's Defence Yearbook 1989* (London: Brassey's, 1989), 67–88; A. H. Cordesman, 'Creating Weapons of Mass Destruction', *Armed Forces Journal International* (Feb. 1989), 54–7; H. J. McGeorge, 'Chemical Addiction', *Defense and Foreign Affairs*, 17/4 (Apr. 1989), 16–19, 32–3; S. R. Tesko, 'It's not just the Soviets anymore', *National Defense* (Apr. 1989), 31 ff.; Senator J. S. McCain, 'Proliferation in the 1990s: Implications for US Policy and Force Planning', *Strategic Review* (summer 1989), 9–20; W. S. Carus, 'The Genie Unleashed: Iraq's Chemical and Biological Weapons Production', *Policy Papers* [Washington Institute for Near East Policy], 14 (1989); W. S. Carus, 'Why Chemical Weapons Proliferate: Military and Political Perceptions', in E. H. Arnett (ed.), *New Technologies for Security and Arms Control: Threats and Promises* (Washington: American Association for the Advancement of Science, 1989), 279–85; T. M. Weekly, 'Proliferation of Chemical Warfare: Challenge to Traditional Restraints', *Parameters: US Army War College Quarterly*, 19/4 (Dec. 1989), 51–66; E. D. Harris, 'Chemical Weapons Proliferation: Current Capabilities and Prospects for Control', in *New Threats: Responding to the Proliferation of Nuclear, Chemical, and Delivery Capabilities in the Third World: An Aspen Strategy Group Report* (University Press of America, 1990), 67–87; A. J. Miller, 'Toward Armageddon: The Proliferation of Unconventional Weapons and Ballistic Missiles in the Middle

Several different strands of motivation can be detected in what was said then, including pure institutional self-interest. Much of the motivation does indeed appear to have been driven by perceptions of indirect threat mediated through the Middle East, but a more immediate menace, too, was seen: tangible direct threats to the security of countries of the North from chemical-warfare weapons spreading among countries of the South. This was a theme with at least four variations, as follows.

First, as Table 4.1 has suggested, it is along the North–South dimension that chemical-warfare weapons have historically displayed their greatest utility. However, rather plausible scenarios— not many, but some—can be envisaged in which, along that North–South dimension, the signs are reversed: scenarios in which, despite technological superiority, the armed forces of an interventionary power are peculiarly vulnerable to chemical-warfare attack. For example, the 'projection of power' into remote regions inevitably stretches lines of communication; some of the technology which might then come to be relied upon heavily is technology that may not have been designed for a toxic environment.[29] Here one may recall that the quantity of mustard gas which disabled the SS *Bisteria* in December 1943, a few hours out of Bari harbour, was probably no more than a kilogram or two.[30] There is new potential, in other words, for that property of

East', *Journal of Strategic Studies* (summer 1990); A. Rathmell, 'Chemical Weapons in the Middle East: Syria, Iraq, Iran, and Libya', *Marine Corps Gazette* (July 1990), 59–67; M. Eisenstadt, '"The Sword of the Arabs": Iraq's Strategic Weapons', *Policy Papers*, 21 (Aug. 1990); R. Vohra, 'Spread of Chemical Weapons and the West Asian Crisis', *Strategic Analysis* (Delhi), 13/7 (Oct. 1990), 852–66; T. Findlay (ed.), *Chemical Weapons and Missile Proliferation* (Boulder, Colo.: Lynne Rienner, 1991); Burck and Flowerree (eds.), *International Handbook on Chemical Weapons Proliferation*; H. J. McGeorge, 'Iraq's Secret Arsenal', *Defense and Foreign Affairs* (Jan.–Feb. 1991), 6–9; H. J. McGeorge, 'The Growing Trend toward Chemical and Biological Weapons Capability', *Defense and Foreign Affairs* (Apr. 1991), 5–7; W. A. Terrill, 'The Chemical Warfare Legacy of the Yemen War', *Comparative Strategy*, 10/2 (Apr.–June 1991), 109–19; J. P. Zanders and E. Remacle (eds.), *Chemical Weapons Proliferation: Policy Issues Pending an International Treaty* (Brussels: VUB Centrum voor Polemologie, May 1991); A. H. Cordesman, *Weapons of Mass Destruction in the Middle East* (London: Brassey's, 1991); K. R. Timmerman, *The Death Lobby: How the West Armed Iraq* (Boston: Houghton Mifflin, 1991).

[29] For elaboration of this rather delicate matter, see R. A. Robinson and N. Polmar, 'Defending Against "the Poor Man's A-Bomb"', *US Naval Institute Proceedings*, 115/2 (Feb. 1989), 100–3.

[30] SIPRI, *Problem of Chemical and Biological Warfare*, i. 97.

chemical-warfare weapons described earlier: force multiplication. The concept here is of chemical-warfare armament, not so much as the 'poor man's atomic bomb', but more as new-age slingshot for David facing Goliath. To the extent that Northern security depends upon capacity to project power into the South, such slingshot might be regarded as directly threatening security. Presumably it was this type of threat perception that inspired the anti-chemical-warfare preparedness of coalition forces during the 1991 Gulf War.

Secondly, as the 1991 Gulf War also showed and the Iran–Iraq War before that, urban populations can be alarmed towards the point of panic by the prospect of chemical-warfare attack delivered by ballistic missiles. One would doubt that cities of southern Europe, for instance, are any less sensitive in this regard than Tehran proved to be in 1988 or Diyarbakir and Tel Aviv three years later.

Thirdly, and related, chemical-warfare weapons have characteristics in addition to their demoralizing potential which can make them seem particularly suitable armament for irregular 'terrorist' force. It does not do to be specific on this subject; for present purposes one need recall only the insidiousness of those chemical-warfare weapons that exploit delayed-effect toxic agents.[31]

Finally, lying beyond chemical-warfare weapons, capable of using much of their technology, is biological weaponry. History shows a clear tendency for vigorous chemical-warfare-weapons-acquisition programmes to spread, sooner or later, into biological weapons.

So, chemical-warfare weapons as defence against 'power projection', as overt or covert armament of overt or covert terrorists, as harbinger of germ warfare: reasons can indeed be discerned for rich industrialized countries to perceive danger to themselves in chemical-warfare armament, despite the waning of East–West tension. And it can be seen, too, why the danger has been described as growing, given the fashion for chemical-warfare armament that was thought to be sweeping the developing world in the wake of the Iran–Iraq War.

Nor is it a threat which menaces only the North. Hence, pre-

---

[31] There is further discussion in J. P. Perry Robinson, 'An Assessment of the Chemical and Biological Warfare Threat to London', *Greater London Area War Risk Study* (Research Report, 22; 1986).

sumably, the grave concern expressed in the Final Declaration of the 1989 Paris Conference (endorsed by all 149 delegations, among them the important states of the Middle East) about the 'growing danger posed to international peace and security by the risk of the use of chemical weapons as long as such weapons remain and are spread'.

## CHEMICAL-WARFARE ARMAMENT IN THE MIDDLE EAST

The present state of that 'spread' of chemical-warfare weapons is difficult to specify. The picture is in shades of grey, not the relatively sharp contrasts that distinguish the 'haves' from the 'have-nots' in the realm of nuclear weapons. Its dominant feature is the diffusion from North to South of manufacturing technologies that are 'dual use': technologies within the broad fields of industrial chemistry and biotechnology, often of great commercial and societal value, that also happen to be applicable in the mass production of chemical-warfare weapons. The requisite raw materials and feedstocks are also frequently 'dual use', and not a few commercial chemicals and civil industrial intermediates are themselves potential chemical-warfare agents. More and more countries are, in consequence, becoming capable of producing chemical-warfare weapons without necessarily wanting to be.

Clearly some states have decided to become chemical-warfare capable in the military sense too. Only Iraq has actually advertised the fact. The rest have preferred to keep quiet about their programmes, maintaining tight security around them. And intelligence gathered by others about those programmes is generally also held secret, for the usual reasons. Although a good many states are no doubt studying the option, the number with actual stockpiles of chemical-warfare weapons that have real military significance is probably not great: in January 1989 the then Director of the US Arms Control and Disarmament Agency, General Burns, told the Senate Foreign Relations Committee that, besides the United States and the Soviet Union, 'no more than a handful, five or six, actually possess a stockpile'.[32] Most of them seem to be in the Middle East.

[32] Major General W. Burns, as quoted in R. J. Smith, 'Lawmakers Plan Chemical Weapons Curb', and 'Agency Gets Last Word on Poison Gas', *Washington Post*, 25 Jan. 1989, A9; 13 Dec. 1989, A23.

It would be a rash person, however, who undertook to describe the true state of chemical-warfare armament in the Middle East or anywhere else. For the reasons just outlined, the appearances are deceptive, the realities ambiguous. The secrets are guarded, and, when some are disclosed, the possible motivations make the information suspect. There is much gossip, and much quoting of gossip as though it were well-founded information. The existing published literature on the subject is, therefore, to put it mildly, unreliable. Parts of it are no doubt accurate, but its readers generally have no *prima facie* way of knowing that. They are acutely vulnerable to disinformation.[33]

However, reliability tests of the literature, at least in regard to information before the 1991 Gulf War about Iraqi chemical-warfare capabilities, are now becoming possible, thanks to the inspections in Iraq of the UN Special Commission that is implementing Section C of Security Council Resolution 687 (1991). So there is some basis for identifying categories of source material which justify respect. With certain exceptions, the academic writings are not showing up well. Only one part of the literature is proving consistently reliable: information placed *attributably* in the public record by *cognizant* officials of the US government. Small though this category is, quite a lot can be learnt from it about proliferation trends over time and about the present situation in the Middle East.

In February 1985 the senior chemical-warfare official in the Office of the Secretary of Defense testified to the US Congress that the 'number of countries having an offensive chemical-warfare program' was rapidly increasing.[34] The testimony, much quoted subsequently, stated the numbers of such countries at different periods over the previous seventy years. It did not identify the countries, however, nor did it say what 'having an offensive chemical-warfare program' actually meant. A context for the testimony is provided in Table 4.3. The table correlates the Defense Department numbers with information about known

---

[33] For elaboration of these points, see J. P. Perry Robinson, 'Chemical Weapons Proliferation: The Problem in Perspective', in Findlay (ed.), *Chemical Weapons and Missile Proliferation*, 19–35, and 'Chemical Weapons Proliferation: Security Risks', in J. P. Zanders and E. Remacle (eds.), *Chemical Weapons Proliferation: Policy Issues Pending an International Treaty* (Brussels: GRIP, 1991), 69–92.

[34] Dr T. J. Welch, Deputy for Chemical Matters, Office of the Assistant to the Secretary of Defense for Atomic Energy, testimony before the Senate Armed Services Committee, 28 Feb. 1985.

TABLE 4.3. *Chemical-warfare-weapons proliferation and deproliferation, 1915–1984*

| Period | 'Countries having an offensive CW program' (US Defense Dept., 1985) | | Countries reliably identified as possessors of poison-gas weapons | Alleged users of poison-gas weapons |
|---|---|---|---|---|
| | No. | % of the world's independent states | | |
| 1915–18 | 8 | 17 | Austro-Hungary France Germany Italy Russia Turkey (supplied by Germany) United Kingdom United States | [all possessors save Turkey used their chemical-warfare weapons during the First World War] |
| 1918–33 | 5 | 8 | — | China France Italy Spain United Kingdom |
| 1933–45 | 13 | 19 | Canada Czechoslovakia France Germany | Germany Italy Japan Poland |

| Period | | | | |
|---|---|---|---|---|
| | | Hungary<br>Italy<br>Japan<br>Netherlands East Indies<br>Poland<br>South Africa<br>Soviet Union<br>United Kingdom<br>United States<br>Yugoslavia | | Soviet Union<br>Spain |
| 1945–60 | 6 | France<br>Soviet Union<br>United Kingdom<br>United States | 6 | China<br>Cuba<br>France<br>Greece<br>Israel<br>Spain<br>Taiwan<br>United Kingdom<br>United States |
| 1960–70 | 7 | France<br>Soviet Union<br>United States | 5 | Egypt<br>Iraq<br>Israel<br>Portugal<br>United States<br>Vietnam, north |

TABLE 4.3. *Continued*

| Period | 'Countries having an offensive CW program' (US Defense Dept., 1985) | | Countries reliably identified as possessors of poison-gas weapons | Alleged users of poison-gas weapons |
|---|---|---|---|---|
| | No. | % of the world's independent states | | |
| 1970–80 | c.13 | 8 | Soviet Union United States | China Laos Morocco Rhodesia South Africa Soviet Union Vietnam Zaïre |
| 1980–84 | c.16 | 10 | Iraq Soviet Union United States | Burma El Salvador Ethiopia Iraq Israel Laos Philippines South Africa Soviet Union Thailand United States Vietnam |

*Source*: Sussex-Harvard Information Bank on Chemical- and Biological-Warfare Armament and Arms Limitation.

and alleged possessors of actual stocks of poison-gas weapons—using, however, only one category of information about alleged possessors, namely the identities of alleged users.

What Table 4.3 illustrates rather clearly is the deproliferation that has been taking place since the Second World War alongside the proliferation emphasized in the testimony. That deproliferation would show up still more clearly were the table to be extended in time—far enough to catch the closure of the Soviet Union's chemical-warfare-weapons programme in 1987 and the final throes of the US binary-munitions programme in 1990.

What of the contemporary situation? In March 1991 the then Director of US Naval Intelligence, Admiral Brooks, spoke in testimony before a Congressional committee of no less than twenty-four countries outside NATO and the Warsaw Pact that 'probably possess offensive chemical-warfare capability' (these he identified as Burma, China, Egypt, India, Iran, Iraq, Israel, Libya, North Korea, Pakistan, South Korea, Syria, Taiwan, and Vietnam) or that 'may possess' such a capability (Indonesia, Saudi Arabia, South Africa, and Thailand), or that are 'developing or suspected of seeking' the capability (these six he did not identify).[35] The *Washington Post* later reported that his lists of countries represented the latest joint assessment of the CIA and Defense Intelligence Agency (though it did not report what was meant by 'capability').[36]

Prominent in those lists are countries of the Middle East. In January and February 1992 this point was emphasized by the incoming Director of Central Intelligence, Robert Gates, during public testimony before several Congressional committees:

Most major Middle Eastern countries have chemical weapons development programs, and some already have weapons that could be used against civilians or poorly defended military targets. Most have not yet equipped their delivery systems to carry weapons of mass destruction. Over the next decade, however, we expect such weapons to become more

[35] Statement of Rear-Admiral T. A. Brooks, USN, Director of Naval Intelligence, before the Seapower, Strategic, and Critical Materials Subcommittee of the House Armed Services Committee, on Intelligence Issues, 7 Mar. 1991, pp. 56–9; 'Navy Report Asserts Many Nations Seek or have Poison Gas', *New York Times*, 10 Mar. 1991, p. 15.

[36] R. J. Smith, 'Confusing Data on Chemical Capability', *Washington Post*, 15 Mar. 1991, A21.

widespread from North Africa through South Asia if international efforts fail to curtail this proliferation.[37]

Director Gates was, it should be remarked, drawing attention more to the future dangers of chemical-warfare armament in the Middle East than to present ones.

As to the nature of current capabilities, both he and his predecessor in office, Judge Webster, have given public testimony, though it does not extend to the chemical-warfare capabilities of Egypt, Israel, and Saudi Arabia referred to by Admiral Brooks. The particulars are as follows.[38]

## Iraq

Production and stockpiling of chemical-warfare agents began in the early 1980s and still continued even after the Iran–Iraq War ceasefire, with several thousand tons having been made by February 1989. Much of the production had been consumed during the Iran–Iraq War. In December 1990 Iraqi military forces had roughly 1,000 tons of chemical-warfare agents on hand. Initially the production programme had been heavily dependent upon the assistance of West European firms, which supplied technology and precursors. That dependence later became much reduced, complete independence being the aim. Mustard gas, tabun, and sarin were produced at Samarra. There were ancillary production facilities elsewhere. The munitions into which the agents were filled included bombs, artillery shell, and rockets. Further particulars, as observed by the UN Special Committee, are given in Table 4.4. Desert Storm was assessed to have destroyed

[37] Statement of the Director of Central Intelligence before the Senate Armed Services Committee, 22 Jan. 1992.

[38] Drawn from W. H. Webster, prepared statement before the Senate Governmental Affairs Committee, 9 Feb. 1989; W. H. Webster, prepared statement before the Senate Foreign Relations Committee, 1 Mar. 1989, hearings, *Chemical and Biological Weapons Threat: The Urgent Need for Remedies*, 29–33; William Webster speaking to *Washington Post* reporters, as reported in G. Lardner, jr., with D. Balz, R. J. Smith, and M. Moore, 'No Iraq Move Seen until Attack Near: CIA Expects Saddam to Extend Crisis', *Washington Post*, 16 Dec. 1990, p. 1; R. Gates, '[prepared written] Proliferation Testimony for Sen. Glenn's Governmental Affairs Committee on 15 Jan. [1992]'; statement of the Director of Central Intelligence before the Senate Armed Services Committee, 22 Jan. 1992; and statement of the Director of Central Intelligence before the House Foreign Affairs Committee, 25 Feb. 1992.

TABLE 4.4. *Chemical weapons in Iraq: An unofficial listing from UN Special Commission*

|  | No. | Agent tons (approx.) |
|---|---|---|
| **Filled chemical munitions** | | |
| Warheads, Al-Hussein missile, sarin[a] | 16 | 2 |
| Warheads, Al-Hussein missile, binary, alcohol-filled[b] | 14 | — |
| Warheads, rocket, 122 mm (several variants), sarin | c.11,000 | 40 |
| Aircraft bombs, R-400, binary, alcohol-filled | 336 | — |
| Aircraft bombs, DB-2, sarin | c.200 | 40 |
| Aircraft bombs, LD-250, mustard gas | 915 | 75 |
| Aircraft bombs, AALD-500, mustard gas | 676 | 100 |
| Artillery projectiles, 155 mm, mustard gas | 12,700 | 60 |
| Mortar projectiles, 82 mm and 120 mm agent CS[c] | c.20,000 | 25 |
| **Unfilled chemical munitions** | more than 79,000 | — |
| **Bulk chemical agent[d]** | | |
| Tabun nerve gas, spoiled | | 50 |
| Sarin nerve gas | | 150 |
| Mustard gas | | over 400 |

[a] Iraqi sarin was made in two versions: straight isopropyl methylphosphonofluoridate (agent GB) and a mixture of that and its O-cyclohexyl homologue (agent GF).

[b] These warheads had a capacity of about 140 litres, and were filled with 70 litres of alcohol. The idea apparently was to tip in 2–3 cans of the sarin-precursor DF prior to launch. The agent-filled warheads had had their DF added just prior to leaving Muthanna.

[c] Other CS munitions have been found, including rocket-propelled grenades for RPG-7 type shoulder-fired launchers, charged with a solution of CS in methylene chloride.

[d] In addition, some 3,800 tons of agent-production intermediates (precursors and precursors of precursors) have been found.

*Source: Chemical Weapons Convention Bulletin*, 13 (Sept. 1991), 14 (Dec. 1991), and 15 (Mar. 1992), *passim*.

much of the production infrastructure, but, as of January 1992, the CIA believed that enough 'production capability' had been preserved to allow production of 'modest quantities of chemical agents' to resume almost immediately; 'a year or more', however, would be needed 'to recover the chemical-warfare capability previously enjoyed' by Iraq.

### Syria

Production of a variety of chemical-warfare agents and munitions, and their stockpiling for battlefield missions, began in the mid-1980s with assistance from West European firms, who were instrumental in supplying the required precursor chemicals and equipment. Some Syrian weapon systems can now deliver nerve gas. By early 1992 Syria appeared to be seeking assistance from China and Western firms to improve its chemical-warfare (or biological-warfare) warhead technology.

### Iran

Production of chemical-warfare agents, including mustard, blood, and nerve gases, began at a factory in the vicinity of Tehran in the mid-1980s, supplied with chemical-processing equipment and precursors by West European and Asian firms. As of February 1989 the programme was still expanding. Agent-filled bombs and artillery shells were then being stockpiled for battlefield missions.

### Libya

As of February 1989 a factory for mustard and nerve gases was being built, with an associated munition-filling plant, at Rabta. It was not then ready for large-scale production and was expected to remain dependent on assistance from West European and Asian firms. Its projected capacity, on the order of tens of agent tons per day, seemed to make it the largest in the Third World, though smaller than the combined Iraqi capacity.[39] Three years later, the

[39] This comparison presumably included the Iraqi production facilities not only at Samarra (Muthanna) but also at Fallujah, where there had been production of DF, the binary-munition fill that is a sarin-precursor. The UN Special Committee has collected information which puts the capacity of the now-defunct Iraqi

CIA estimate was that one hundred tons of chemical agents had been produced and stockpiled at the Rabta facility before cleaning-up operations, 'perhaps in preparation for the long-awaited public opening of the facility to demonstrate its alleged function of producing legitimate pharmaceuticals'.

## CONCLUSION

Chemical-warfare weapons have been in the Middle East for many decades now. Their military significance there has not been substantial, except possibly—though not yet demonstrably—under the special circumstances that characterized the final years of the Iran–Iraq War. However, self-interested issue-creation in the West on the theme of chemical proliferation has enhanced the political significance of chemical-warfare weapons in the Middle East. This factor, amplified by trends of a technological nature (such as the spread of guided-missile technologies), may conceivably carry chemical-warfare armament forward into an altogether greater salience upon regional security. In that event, armed forces in the region may become compelled to adapt themselves more closely to the various special military potentials which chemical-warfare armament possesses.

But such assimilation would run counter to the experience of past possessors of chemical-warfare weapons, such as the United Kingdom and France. After some initial enthusiasm, these states became content in the end to abandon chemical-warfare weapons, evidently concluding that the military potential of the weapons was insufficient to justify the opportunity and other costs of the adaptation needed to be able to exploit them.

In the particular security environment of the Middle East, that historical experience might or might not repeat itself. If, to take the pessimistic view, it does not, the chemical-warfare weapons themselves would probably not become especially threatening or destabilizing, for they would certainly induce greater attention to antichemical protection. The real danger would lie rather in the

mustard-gas factory at 5 tons/day and that of the sarin/GF factory at 2.5 tons/day; see *Chemical Weapons Convention Bulletin*, 13 (Sept. 1991), 11. It is instructive to recall that the former US sarin factory, at Rocky Mountain Arsenal near Denver, had a capacity of about 100 tons/day, while the capacity of the German mustard-gas factory at Gendorf during the Second World War was some 130 tons/day.

indirect consequences, above all in the boost that would be given to biological-warfare armament.[40]

This, one may very well conclude, is a predicament for which arms control, provided it does not inhibit antichemical protection, is the obvious remedy. The CWC, coupled with the regional contacts that have already been initiated, means that it is a remedy within reach.

[40] Egypt, Israel, and Syria have all kept themselves outside the disarmament regime established by the 1972 BTWC; Iraq has joined it only under duress.

# 5

# BIOLOGICAL WEAPONS: THEIR NATURE AND ARMS CONTROL

## GRAHAM S. PEARSON

THE term 'biological weapons' is little understood and little appreciated by the public at large. Whilst there is some comprehension of chemical weapons, largely arising from the use of such weapons in the First World War, and the lasting images of incapacitated troops in that war, together with the use by Iraq of chemical weapons against Iran and against its own Kurdish population in the mid-1980s, there is no such general appreciation of biological weapons or biological warfare. All too often, biological warfare conjures up images of uncontrollable epidemics attacking both aggressor and attacked. This is not the case. Although transmissible biological-warfare agents could be selected by an aggressor, there are many agents that are non-transmissible and will affect only the target population, provided the aggressor takes care to deliver the agent in such a way that his forces are not exposed, or are protected, either by immunization of the body by vaccination or by physical protection using respirators. Additionally, by selection of the biological-warfare agent used, the outcome for the attacked population can be incapacitation or death.

### INTRODUCTION

The revulsion against the use of chemical weapons in the First World War led to the Geneva Protocol of 1925, which prohibited the use of chemical weapons and was extended to include a ban on the use of biological (bacteriological) weapons: 'That the High Contracting Parties, so far as they are not already Parties to Treaties prohibiting such use, accept this prohibition, agree to extend this prohibition to the use of bacterial methods of warfare and agree to be bound as between themselves according to the terms of this declaration.'

Although the 1925 Geneva Protocol has now been signed by over 125 states parties, some thirty of the major nations which signed the Protocol have done so with a reservation. Thus the United Kingdom signed with the following reservation:

a. The said Protocol is only binding on His Britannic Majesty as regards those Powers and States which have both signed and ratified the Protocol or have finally acceded thereto.

b. The said Protocol shall cease to be binding on His Britannic Majesty towards any Power at enmity with Him whose Armed Forces, or the Armed Forces of whose allies, fail to respect the prohibitions laid down in the Protocol.

In September 1991, at the Third Review Conference of the BTWC, the United Kingdom withdrew its reservation to the Geneva Protocol in respect of biological weapons.

## UK Policy

Although the United Kingdom developed a retaliatory chemical-warfare capability in the First World War and maintained this throughout the Second World War, it only investigated a retaliatory biological-warfare capability during the Second World War and thereafter into the 1950s. Even though biological warfare was recognized as having both strategic and tactical military utility, the decision was taken in the late 1950s to abandon offensive chemical and biological warfare. Consequently, since that time, the United Kingdom has been solely concerned with the provision of effective protective measures for the UK Armed Forces against the threat that chemical or biological weapons might be used against them by an aggressor. A key strand in UK policy has been to take a leading role in chemical and biological arms control, and the United Kingdom is a co-depository along with the United States and the Soviet Union of the BTWC of 1972.

## Definitions

What exactly are biological weapons—and how do they differ from chemical weapons? The answer lies in the nature of biological-warfare agents and in their mode of action. However, as is so often the case, there is a grey area between chemical and biological warfare. It is clear that there is a potential spectrum of chemical- and biological-warfare agents ranging from the classical chemical-warfare agents such as mustard, hydrogen cyanide, and the nerve agents through emerging chemical-warfare agents such as toxic industrial, pharmaceutical, or agricultural chemicals to bioregulators and toxins, which are the products of living materials, to genetically manipulated micro-organisms, and thus to the naturally existing and traditional biological-warfare agents such as anthrax, tularemia, and plague. The spectrum is shown in Fig. 5.1. The scientific division between chemical and biological weapons is that chemical weapons are non-living chemicals (left-hand four boxes) which poison the target population, whilst biological weapons are living micro-organisms (right-hand two boxes) which infect the target population. Potential confusion arises on two counts:

1. The BTWC of 1972 applies to biological agents and toxins. Neither of these is defined in the treaty. It is, however, evident that the term 'biological agents' applies only to the microbial organisms that are living and are able to replicate themselves and infect the target system. The toxins, on the other hand, although the natural products of microbial organisms or plants, are non-living and are strictly chemicals.

2. The term 'agents of biological origin' is frequently used but has no standing in respect of the BTWC or other treaty negotiations. 'Agents of biological origin' can cover a wider range of agents and, presumably, covers any material that is produced or can be produced by biological systems. It therefore embraces both non-living chemicals and living micro-organisms.

For this chapter, the term 'biological weapons' is taken to include toxins, as it is this definition that is the subject of the BTWC of 1972.

Another facet of biological warfare requires clarification. What is the target of biological warfare and what is the route of attack? There is general agreement that biological weapons may be used to attack human beings, animals, or plants. Proposals were made

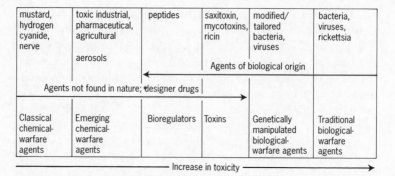

| mustard, hydrogen cyanide, nerve | toxic industrial, pharmaceutical, agricultural aerosols | peptides | saxitoxin, mycotoxins, ricin | modified/ tailored bacteria, viruses | bacteria, viruses, rickettsia |
|---|---|---|---|---|---|
| | | | | Agents of biological origin | |
| Agents not found in nature; designer drugs | | | | | |
| Classical chemical- warfare agents | Emerging chemical- warfare agents | Bioregulators | Toxins | Genetically manipulated biological- warfare agents | Traditional biological- warfare agents |

Increase in toxicity ⟶

FIG. 5.1. The chemical- and biological-warfare spectrum

that the 1991 Third Review Conference should affirm this in its declaration; in contrast, chemical weapons are seen as materials that are harmful to animals and to human beings; use against plants is unclear. As to the route of primary attack of the target by biological weapons, this is varied:

1. Inhalation. This is the main route of attack of human beings. The biological-warfare agent is retained within the respiratory system and the lungs.
2. Ingestion of contaminated food and water. This presents difficulties for a large-scale attack.
3. Contamination of an open wound with biological-warfare agent. For obvious reasons, this is not a particularly effective route of attack. Contaminated bullets and fragmentation weapons are proscribed by several Hague Conventions.
4. Insect vectors. Infected insects can transmit disease to human beings.

Most of the traditional biological-warfare agents are non-transmissible, and consequently will not be transmitted from those exposed to the attack to those not exposed. However, if a transmissible agent is selected, then the disease can be transmitted between human beings.

### HISTORICAL PERSPECTIVES

Allegations of biological warfare go back to ancient times. Table 5.1 includes some of the conflicts in which biological warfare is alleged to have been used.

Historically, BC 190 saw Hannibal winning a naval victory by firing earthen vessels full of venomous snakes into the ships of the enemy. In the eighteenth century the Russians threw bodies of plague victims into Swedish cities, the British used smallpox-

TABLE 5.1. *Milestones in chemical and biological warfare*

| Period | Incident |
| --- | --- |
| BC 673 | 'Greek fire' at Siege of Constantinople Contamination of water supplies |
| BC 600–200 | Greek use of smoke |
| BC 190 | Hannibal's naval victory using venomous snakes |
| 1710 | Russians throw bodies of plague victims into Swedish city |
| 1763 | British use of smallpox-contaminated blankets (Fort Pitt, Ohio) |
| 1797 | Napoleon's attempt (Italian campaign) to infect with swamp fever |
| First World War | Large-scale use of chemical warfare |
| 1925 | Geneva Protocol |
| 1935–8 | Ethiopia |
| Second World War | Unit 731, Japan |
| 1951–3 | North Korea and China allegations |
| 1957 | Oman allegations |
| 1960s | Vietnam War allegations |
| 1963–7 | Yemen |
| 1972 | BTWC |
| 1979 | Sverdlovsk anthrax outbreak |
| 1980s | Yellow rain in South-East Asia |
| 1980s | Afghanistan |
| 1980s | Middle East requests for cultures |
| 1985–8 | Iran–Iraq war |

contaminated blankets (Fort Pitt, Ohio), and Napoleon in his Italian campaign attempted to infect the inhabitants of the besieged city of Manchua with swamp fever. The next significant mention of biological warfare comes in the Second World War with the activities of Unit 731 in Japan. Although there is a paucity of definitive information, it appears certain that Japan used prisoners of war to evaluate the effects of various candidate biological-warfare agents. There are also allegations of use of biological warfare by Japan against China in the 1930s.

Another landmark was the use of Gruinard Island in 1942–3 to carry out trials to determine whether anthrax spores could be disseminated from bombs and whether such spores borne down-wind would infect a flock of sheep. The experiments showed that this could be done and Gruinard remained a prohibited place until decontaminated in 1986 and returned to its owner in 1990. It is, however, important to recognize that the actual hazard from anthrax spores on Gruinard Island during the intervening years was slight. In the early years, the extent of the contamination was unknown. In reality, the risk of sufficient anthrax spores being re-aerosolized to present a hazard was minimal. As the island was to be returned to its original owners, the Ministry of Defence considered that all traces of anthrax should be destroyed, which led to the island being decontaminated in 1986 using treatment with a 5-per cent solution of formaldehyde in sea water, leading to it being declared safe for man and beast and returned in 1990 to its original owners.[1]

The agreement of the BTWC in 1972, prohibiting the development, production, and stockpiling of biological and toxin weapons, represented the next milestone. This convention is discussed at some length below.

The next milestone in the story of biological warfare comes with the outbreak of anthrax in the Soviet city of Sverdlovsk. Early in 1980, articles appeared in various Western newspapers claiming that in April 1979 there had been an explosion at 'Military Village 19' in the Sverdlovsk area in which anthrax bacteria had been released and resulted in a number of deaths

[1] R. J. Manchee and W. D. P. Stewart, in *Chemistry in Britain*, (July 1988), 690.

reported as ranging from three hundred to a thousand. The US State Department asked the Soviet Union for an explanation in March 1980, which produced a reply that an outbreak of Siberian fever (i.e. anthrax) had indeed occurred in March–April 1979 in the Sverdlovsk area, but that it had been caused by the illegal sale of infected meat. The US State Department issued a statement later in March, indicating that it did not consider the Soviet response plausible and remained concerned about the incident, as the United States had ruled out natural causes as a likely explanation.

Subsequent Press articles continued to express concern and to report further details about the incident, including statements that a large-scale clean-up operation had occurred. Experiments performed to simulate accidents which may occur during large-scale production of micro-organisms were carried out by the Chemical Defence Establishment.[2] Four types of accident chosen to be the most likely to result in the greatest hazard to health were simulated using a bacterial model; all four accidents were concerned with faults occurring in the operation of a fermenter. The results showed that, in one experiment, no aerosol was produced; in the other three, the fraction of expelled bacteria from the fermenter in the form of an aerosol was between $1 \times 10^{-3}$ to $4 \times 10^{-5}$ per cent. The respiratory hazard resulting from this aerosol was very much less than that which would have resulted from an intentional aerosolization of the contents of the fermenter.

Further discussions on the Sverdlovsk incident occurred in 1986 at the Second Review Conference of the BTWC; the first reports about Sverdlovsk had appeared in March 1980 at the same time as the First Review Conference. In 1986 the Soviet Union at the Review Conference claimed that the anthrax outbreak was due to the sale of contaminated meat on the black market and provided additional medical details. In 1988 three Soviet public-health officials visited the United States to expand on their explanation. They stated that unsterilized, anthrax-contaminated bone-meal was fed to cattle, and some became infected. Those that fell ill were killed and their meat sold. Ninety-six people became infected

---

[2] J. Ashcroft and N. P. Pomeroy, in *Journal of Hygiene Cambridge*, 91 (1983), 81.

after eating the contaminated meat, seventy-nine developed the gastric form of anthrax, and sixty-four died. None had lung infections. According to the Soviet health officials, all ninety-six were civilians and mainly male, and only one child was infected. The discussions were limited to scientific questions about the medical information presented; other questions on locations and districts were refused. The matter rested with no satisfactory explanation.

Then, in August 1990, a major state newspaper published an article entitled 'Accident at Sverdlovsk'.[3] This described an accident in April 1979 in which people died of anthrax. The memories of the doctors concerned, that those who died did so from inhalation anthrax, is convincingly reported. The article is highly critical of the paper by Professors Bezdenezhayy and Nikiforov, which makes no mention of the inhalation form of anthrax.[4] The author of the 1990 article concludes by asking three questions:

a. Why did people in military uniforms play a prominent role in the examination of facts, if nothing but a few cases of anthrax occurred at Sverdlovsk?

b. Why were all records, all case histories, and the entire documentation destroyed in all the establishments involved in the case?

c. And, finally, if there had been nothing else, why were people authorized to check things up here and there eleven years after the event?

and calls, on behalf of the editorial board of the *Literaturnaya Gazeta*, for the Parliament to set up a board of enquiry into the affair. More recently, in October 1991, the *Wall Street Journal* published a series of articles on the Sverdlovsk incident,[5] and in November and December articles appeared in *Izvestiya*. The overall thrust of these 1991 articles has been to confirm the links to the military village 19 and the claims that the outbreak of anthrax was due to inhalation of spores.

The 1980s also saw the reporting of yellow rain in south-east Asia. Although attacks had been taking place since 1975, the matter did not achieve prominence until the then US Secretary of

[3] Natal'ya Zenova, *Literaturnaya Gazeta*, 34 (22 Aug. 1990).
[4] I. S. Bezdenezhayy and V. Nikiforov, 'Zhurnal Mikrobiologii', *Epidemiologii i Immunobiologii*, 5 (1980), 111.
[5] *Wall Street Journal*, 21 Oct. 1991.

State, Alexander Haig, sent a report (Special Report Number 98) to Congress entitled 'Chemical Warfare in South East Asia and Afghanistan', dated 22 March 1982.[6] This was followed by a further report providing an update from the US Secretary of State, George P. Shultz, in November 1982 (Special Report Number 104).[7] The report presented the key judgement that:

Selected Laos and Vietnamese forces, under direct Soviet supervision, have employed lethal trichothecene toxins and other combinations of chemical agents against H'Mong resisting Government control and their villages since at least 1976. Trichothecene toxins have been positively identified, but medical symptoms indicate that irritants, incapacitants and nerve agents have also been employed. Thousands have been killed or severely injured. Thousands also have been driven from their homeland by the use of these agents.

There was much debate in the academic community about the validity of these claims. An article in the *Scientific American* of September 1985 concludes that yellow rain was the faeces of honey bees and was not an agent of chemical warfare.[8] A later article in the spring 1987 issue of *International Security* concludes that the evidence available in the public domain simply does not support the yellow rain–trichothecene mycotoxin views.[9] In so far as the United Kingdom is concerned, a parliamentary answer in May 1986 stated that scientists at the Chemical Defence Establishment had not found trichothecene mycotoxins in any samples of yellow rain from South-East Asia.[10] Nevertheless, the UK Government believed on the basis of epidemiological evidence that chemical-warfare attacks probably did take place in south-east Asia, although it was unable to identify the agent or agents used

[6] Special Report No. 98, 'Chemical Warfare in South-East Asia and Afghanistan', Report to the Congress from Secretary of State Alexander M. Haig, jr., 22 Mar. 1982.
[7] Special Report No. 104, 'Chemical Warfare in South-East Asia and Afghanistan: An Update', Report from Secretary of State George P. Schultz, Nov. 1982.
[8] T. D. Seeley, J. W. Nowicke, M. Meselson, J. Guillemin, and P. Akratanakul, in *Scientific American*, 253/3 (1985), 128.
[9] E. D. Harris, 'Sverdlovsk and Yellow Rain: Two Cases of Soviet Non-compliance', *International Security*, 11/4 (1987), 41.
[10] Hansard, *Commons*, 98 (19 May 1986), Col. 92.

or to say for certain who may have supplied them. The work carried out by the Chemical and Biological Defence Establishment and subsequently published indicates the necessity of obtaining and analysing samples immediately after any alleged attack; samples from alleged victims are particularly difficult to analyse, as the body is extremely efficient in metabolizing any chemicals whether from a chemical- or toxin-weapon attack, or a drug used to treat the symptoms of any illness.

Over the past few years there have been various allegations of the use of biological warfare. Many of these allegations are unsubstantiated or unproven. Why such uncertainty? Therein lies one of the subtleties of biological warfare, in that use does not necessarily produce a clear signature. This is especially true if the selected biological-warfare agent produces a disease that is endemic in the country whose population is being attacked. Natural outbreaks of endemic disease occur from time to time, such as the outbreaks of Q-Fever in the United Kingdom in Birmingham in May 1989 and in Northern Ireland at Ballycastle in the same month. Moreover,

TABLE 5.2. *Active-duty US Army patients*

| Theatre | Date | Patients suffering from | |
|---------|------|-------------------------|---|
| | | Disease and non-combat injury (%) | Battle injury (%) |
| Second World War (Pacific) | Nov. 1942–Aug. 1945 | 95 | 5 |
| Second World War (European) | June 1944–May 1945 | 77 | 23 |
| Korean War | July 1950–Dec. 1969 | 83 | 17 |
| Vietnam War | Jan. 1969–Dec. 1969 | 82 | 18 |
| | Jan. 1967–Dec. 1967 | 76 (disease) 14 (non-combat) | 16 |

*Source*: Adapted from publications from the Medical Department of the US Army Office of the Surgeon General, Centre of Military History.

in the event of conflict, it is vital to recognize that disease is rampant; Table 5.2 gives data for the occurrence of disease in recent hostilities which make it clear that the majority of casualties in military hospitals were suffering from disease; the proportion of battle casualties in the Vietnam War in 1967 was less than 1 in 7.[11] Although some biological-warfare agents can produce an unusual disease for the locality in which they are used, it seems probable that an aggressor would choose to use an agent which might occur naturally, and consequently, not be readily recognized as such against a background of naturally occurring disease.

### BIOLOGICAL-WARFARE AGENTS AND DELIVERY MEANS

The range of potential biological-warfare agents has been recognized for many years: two publications in 1969—one by the United Nations[12] and one by the World Health Organization[13]—provide a list of typical candidate agents indicating their potency and the availability of vaccines against the disease. The main groups or classes of biological-warfare agents are:

1. *Bacteria*. These are the causative micro-organisms that produce diseases such as anthrax, plague, and tularemia. Although many pathogenic bacteria are susceptible to antibiotic drugs, strains can be selected that are antibiotic resistant and occur naturally. They can be readily grown in artificial media using facilities akin to those in the brewery industry.
2. *Viruses*. There are large numbers of viruses that might be potential biological-warfare agents producing diseases such as Venezuelan equine encephalitis. These are the smallest forms of life and must be grown on living tissue.
3. *Rickettsia*. An example is the organism that produces Q-

---

[11] Information from publications from the Medical Department of the US Army Office of the Surgeon General, Centre of Military History.
[12] Report of the UN Secretary General, 'Chemical and Bacteriological (Biological) Weapons and the Effects of their Possible Use', A/7575/Rev 1, S/9292/Rev 1 (1969).
[13] *Health Aspects of Chemical and Biological Weapons* (Geneva: WHO, 1970).

Fever. These are intermediate between viruses and bacteria and must be grown in living tissue.
4. *Fungi.* An example is coccidioidomycosis. Relatively few species appear to have biological-warfare potential.
5. *Toxins.* These are the non-living products of micro-organisms such as botulinum toxin or staphylococcal enterotoxin B, of plants such as ricin from castor beans or of living creatures such as saxitoxin from shellfish.

Biological-warfare agents vary considerably both in the quantities needed to produce disease or to intoxicate man, and in the nature of the effect—which may be to incapacitate or to kill. The time to effect is of the order of twelve hours or more for toxins, and a few days or more for microbial agents, as the micro-organism has to replicate within the target individual and cause the corresponding disease. This delayed onset of symptoms is one of the main differences between biological and chemical warfare: the latter tends to be faster acting, with nerve agents and hydrogen cyanide producing effects in minutes, although some chemical-warfare agents, such as mustard and phosgene take several hours to produce symptoms. Other differences between chemical and biological weapons are shown in Table 5.3.

The choice of a potential biological-warfare agent involves consideration of a large number of factors—such as the infective dose, the time to effect, and whether the agent produces a transmissible disease (see Table 5.4), as well as the method of attack of the target population (inhalation, ingestion, or by an insect vector), the means of dispersion or delivery of the agent, the stability of the agent, and the practicality of achieving an infective dose at the target personnel.

Similar considerations apply to toxins. However, as toxins are non-living, they cannot produce a transmissible disease. Toxins and other chemical-warfare agents, unlike some biological-warfare agents, can affect only those exposed to the agent; the contact hazard associated with some chemical-warfare agents depends on the physical transfer of enough agent from an individual or a piece of equipment or terrain to another individual to produce effects and is totally different from the transmission of an infectious disease. And only some biological-warfare agents produce transmissible diseases.

TABLE 5.3. *Differences between chemical and biological weapons*

| Parameter | Chemical weapons | Biological weapons |
|---|---|---|
| Potency | milligrams | micrograms (toxins) picograms (micro-organisms) |
| Time to effect | minutes to hours | hours to days |
| Nature of effect | lethality short-term incapacitation (hours) (mustard, phosgene longer) | lethality long-term incapacitation (days, weeks) |
| Specificity | less specificity | greater host specificity |
| Persistence of hazards | limited (mustard, some nerve agent exceptions) | limited (little hazard from re-aerosolization) |
| Signature | short time to effect means clear signature | long time to effect means minimal signature, especially if naturally occurring disease selected |

## Delivery Means

The choice of delivery means depends on the route of attack of the target population. In the Second World War, when the United Kingdom considered how it would retaliate in kind were biological weapons to be used against UK forces, the approach selected was to produce cattle cakes containing anthrax spores which would be disseminated through the flare shutes of aircraft over Germany and then be consumed by cattle which would develop anthrax and die. To be effective, dissemination would have to take place over fields in which cattle were or would be grazing. This limited UK capability to retaliate in kind was destroyed shortly after the Second World War ended.

Other delivery means have been devised that are capable of producing aerosols of the particle size necessary to enter and be retained in the human respiratory system and lungs.

TABLE 5.4. *Biological-warfare agent characteristics*

| Agent | Dose to effect[a] | Time to effect | Mortality | Transmissible |
|---|---|---|---|---|
| Bacteria | | | | |
| anthrax | 20,000 | 1–5 days | fatal | negligible |
| plague | 3,000 | 2–5 days | fatal | high |
| tularemia | >25 | 1–10 days | low | negligible |
| brucellosis | 1,300 | 1–3 weeks | low | none |
| Viruses | | | | |
| VEE (Venezuelan equine encephalitis) | 25 | 2–5 days | low | none |
| Rickettsia | | | | |
| Q-Fever | 1 | 10–21 days | low | none |
| Fungi | | | | |
| coccidioidomycosis | 1,350 | 1–3 weeks | low | none |
| Toxins | | | | |
| botulinum toxin | 0.12 µg | ½–3 days | fatal | none |
| staphylococcal enterotoxin B | 0.1 µg/kg iv | 1 hour | low | none |
| ricin | 0.1 µg/kg im | 1–3 days | fatal | none |

[a] The units are in viable cells for bacteria and fungi and in infectious units for viruses and rickettsia. The dose to effect is by the aerosol route unless otherwise stated.

*Sources:* R. Clarke, *We All Fall Down* (Harmondsworth: Penguin, 1968); SIPRI, *The Problem of Chemical and Biological Warfare* ii. *CB Weapons Today*. (Stockholm: Almqvist & Wiksell 1973); J. Sugiyama and E. M. McKissie, 'Lencocyte Response in Monkeys Challenged with Staphylococcal Enterotoxin'. *J. Bact.* 92 (1966), 349–52; J. Crompton and D. Gall, 'Georgi Markov: Death in a Pellet', *Medico-Legal Journal*, 48/2 (1980), 51–62.

The utility or perceived utility of any form of warfare depends on the perceived advantages and disadvantages. For biological warfare, advantages include the extreme potency, so that small 'quantities are necessary; the lack of signature, if an endemic disease agent is selected; the ease of covert production using dual-purpose facilities; lack of collateral damage; lack of risk to one's own forces, if a non-transmissible agent is selected; and large-scale/strategic, small-scale/tactical use or covert use. The disadvantages of biological warfare include the international opprobrium, as it would be in breach of the Geneva Protocol (1925) and BTWC (1972); the fragility of micro-organisms in the atmosphere, which can limit their effectiveness; the limited effectiveness if prevailing weather conditions are unfavourable; the difficulties and dangers of handling biological-warfare agents; and the uncertainty of effectiveness, as there is no proven prior use in war.

The potential utility of biological warfare is clearly indicated in the UN publication of 1969,[14] and has since been reiterated by Steve Fetter in his article in *International Security* in the summer of 1991.[15] Particles of a simulant disseminated from a ship offshore travelled 750 kilometres downwind, giving very substantial area coverage. Micro-organisms in such an aerosol will decay in sunlight and the atmosphere, reducing the effectiveness of the agent coverage significantly according to the organism selected, but in most cases the area affected would be larger and extend much further downwind than a nerve agent released under the same conditions. Table 5.5 compares the effects of attacks on unprotected population by nuclear, chemical, or biological weapons.[16] Steve Fetter says in his article that the lack of destructive power of conventional weapons and the difficulty of developing nuclear weapons means that chemical and biological weapons will soon be the warhead of choice of emerging states possessing ballistic missiles.[17] I cannot comment on what basis he may have had for that statement, but certainly there can be little

---

[14] UN Report, 'Chemical and Bacteriological (Biological) Weapons'.
[15] S. Fetter, 'Ballistic Missiles and Weapons of Mass Destruction', *International Security*, 16/1 (1991), 5.
[16] UN Report, 'Chemical and Bacteriological (Biological) Weapons'.
[17] Fetter, 'Ballistic Missiles and Weapons of Mass Destruction'.

TABLE 5.5. *Comparative estimates of attacks on unprotected population using a nuclear, chemical, or biological weapon*

| Criteria for estimate | Type of weapon | | |
|---|---|---|---|
| | Nuclear (one megaton) | Chemical (15 tons) | Biological (10 tons) |
| Area affected | up to 300 km² | up to 60 km² | up to 100,000 km² |
| Time delay before effect | seconds | minutes | days |
| Damage to structures | destruction over an area of 100 km² | none | none |
| Normal use after attack | 3–6 months after attack | limited during period of contamination | after end of incubation period or subsidence of epidemic |

*Source*: United Nations, *CBW and the Effects of their Possible Use* (Geneva, 1969).

doubt that the potential strategic capability of biological weapons is comparable to that of nuclear weapons.

### BIOLOGICAL-WARFARE DEFENCE STRATEGY

The provision of effective chemical and biological defence rests on a number of interrelated activities. The first essential element is the assessment and evaluation of the hazard. This comprises several strands:

1. *Identification of the potential chemical- or biological-warfare agent*. The particular substance needs to be identified so that the hazard can be evaluated. This ensures that scarce resources are not devoted to providing protective measures against a substance that presents little hazard.
2. *Evaluation of the infectivity or toxicity of this potential agent*. Such evaluation involves the extrapolation of the

hazard to man and takes into account such factors as breathing rate and stress. Materials increasingly need to be evaluated for both lethality and incapacitation.

3. *Potential military utility to the aggressor.* This evaluation addresses such aspects as whether the potential agent could be produced in sufficient quantity and be weaponized by the aggressor; also whether his perceived delivery means would enable the aggressor to deliver a sufficient quantity to produce harmful effects against the defender's forces. Such evaluations need to be comparative, because a chemical- or biological-warfare agent is only likely to be utilized by the aggressor should it enable him to achieve his particular military objectives more effectively than through the use of conventional weapons.

Any evaluation of the hazard must take into account all of these strands if a realistic assessment and evaluation is to be achieved.

Once the hazard has been assessed, effective protective measures are then based on a number of approaches:

1. *The provision of advice to the Armed Services.* Their operational tactics may have to be modified to minimize the potential hazard.
2. *Detection.* The provision of a range of detection and warning devices which will sense the presence or approach of a harmful concentration of a chemical- or biological-warfare agent and thereby alert the Armed Forces to don their protective measures.
3. *Protection.* This falls into three categories:
    (a) *Respiratory protection.* In general, the most vulnerable part of the body is the respiratory system and the lungs. The provision of a respirator to be worn whenever a hazardous concentration of an agent is in the vicinity is an effective protective measure. The S10 respirator has now entered UK service.
    (b) *Body protection.* Some chemical- and biological-warfare agents, but not all, have a percutaneous effect and harm the body through attack of the skin. Consequently, body protection is required and is provided by an NBC suit which is designed to minimize the

physiological load on the body and to maximize the protection against chemical- and biological-warfare agents. Most current suits are multi-layer and porous. The No. 1 Mk 4 suit has now entered UK service.

(c) *Collective protection.* This is the provision of protection for groups of personnel, either within a protected building (i.e. hardened collective protection) or within a more temporary structure that does not provide protection against shrapnel and fragments (unhardened collective protection). In collective protection, the incoming air supply is filtered to remove any chemical- or biological-warfare agent and provision is made for airlocks for access and exit. Collective protection is provided in most ships, armoured fighting vehicles, and pilot facilities at air bases. It is also required for military medical facilities and to enable rest and relief of services personnel.

4. *Contamination Monitoring.* Once protective measures have been taken, it is necessary to monitor the level of the hazard, so that the protective measures such as the respirator and suit can be removed once it is safe to do so, thereby reducing the physiological stress imposed through the wearing of full protective clothing.

5. *Contamination Management.* Some chemical- and biological-warfare agents are highly persistent and present a prolonged hazard. The disadvantages caused by such agents can be minimized through chemical hardening (i.e. the design of military equipment so as to minimize the presence of crevices and cracks into which an agent may find its way and be retained) and by decontamination.

6. *Medical Countermeasures.* Medical countermeasures need to be provided for those personnel who have been exposed to a chemical- or biological-warfare agent. Medical countermeasures fall into two groups:

(a) *Prophylaxis or pretreatment.* This involves protecting the body in advance of any exposure to a chemical- or biological-warfare agent.

(b) *Therapy.* This involves treatment, after exposure to a chemical- or biological-warfare agent has occurred.

Against the potential chemical- and biological-warfare spectrum, the thrust of current work is to provide broad-band protective measures that will protect against the entire spectrum or against part thereof. Some protective measures are highly effective against a range of the spectrum, whilst others are, currently, agent-specific.

## DEFENSIVE AND OFFENSIVE PROGRAMMES

Claims are sometimes made that programmes to defend the Armed Forces are intended to provide an offensive capability. Such arguments are not well based, even though there is necessarily some commonality in basic research. Whilst assessment of the hazard and the definition of the technical requirements for protective measures necessitate an appreciation of the characteristics and stability of the potential biological-warfare agents, the depth of understanding and the scale of work has to be much greater to support an offensive programme.

The broad differences between defensive and offensive programmes are indicated in Fig. 5.2, which follows the schematic published by David Huxsoll and his colleagues.[18] The key elements for an offensive programme are:

1. studies into the enhancement and maintenance of potency;
2. large-scale production of biological-warfare agents—or the availability of large-scale production facilities (even though such possession is not in breach of the BTWC);
3. large-scale dissemination trials of biological-warfare agents or simulants;
4. studies of and work on delivery systems.

There is a potential difficulty in respect of what represents large scale, as this will necessarily reflect the potential scale of military operations by the state concerned. A small-scale capability might suffice to give a small state a significant military capability against its neighbours. The difficulty is how to recognize those states whose biological production capacity is disproportionate to their

[18] D. L. Huxsoll, C. D. Parrott, and W. C. Patrick III, 'Medicine in Defence', *Journal of American Medical Associates*, 262/5 (1989), 677.

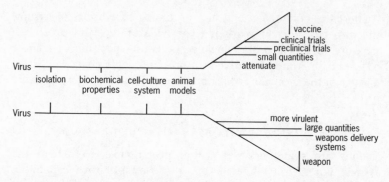

FIG. 5.2. Defensive and offensive biological-warfare programmes (after Huxsoll *et al.*, 'Medicine in Defence')

public-health requirements. The aim for effective arms and export controls is to devise measures that will constrain and block offensive programmes without impairing legitimate defensive and public-health programmes. The difficulties in devising effective measures to control work on biological-warfare agents that give rise to endemic diseases in the region will be evident.

## ARMS CONTROL

### *The Geneva Protocol*

Mention has already been made of the 1925 Geneva Protocol agreed in 1925 following the First World War and which bans the use in war of asphyxiating, poisonous, or other gases and of bacteriological methods of warfare. Because of reservations entered by several of the major nations, the Geneva Protocol is essentially a prohibition of first use of chemical and biological weapons.

The Geneva Protocol makes *no* provision for inspection or verification of allegations of use. In 1989 France held an international conference to reinforce the importance of the 1925 Geneva Protocol. This resulted in a few more nations becoming signatories and a final declaration which emphasized the importance of investigations sponsored by the UN Secretary General. However, the weakness is that such investigations cannot be mounted unless the state in whose territory the alleged attack has taken place is prepared to accept such an investigating team. The refusal to accept such investigations has prevented allegations of use, such as of yellow rain in South-East Asia and of chemical weapons by Iraq against the Kurds being investigated. There is a clear need for an automaticity of access following any allegations of use, whether of chemical or biological weapons. Such access needs to be immediate, as the availability of irrefutable evidence of use of chemical or biological weapons decreases with the passage of time: in particular, the body is very effective in metabolizing and breaking down traces of chemical-warfare agents.

### The Biological and Toxin Weapons Convention

Although bacteriological weapons were included in the 1925 Geneva Protocol, a more comprehensive treaty to prohibit the development, production, and stockpiling of bacteriological (biological) and toxin weapons and their destruction was signed in London, Moscow, and Washington on 10 April 1972. This is often referred to as the Biological Weapons Convention (BWC), but it explicitly includes both living micro-organisms (the right-hand two boxes of the spectrum—see Fig. 5.1) and toxins; its correct full title is 'Convention on Biological and Toxin Weapons'. Article I of the BTWC states:

Each State Party to this Convention undertakes never in any circumstances to develop, produce, stockpile or otherwise acquire or retain:

1. Microbial or other biological agents, or toxins, whatever their origin or method of production, of types and in quantities that have no justification for prophylactic, protection or other peaceful purposes;
2. Weapons, equipment or means of delivery designed to use such agents or toxins for hostile purposes or in armed conflict.

There is no definition of what is intended by microbial or other biological agents, nor is there a definition of toxins. Furthermore, whilst states parties have all agreed not to develop, produce, stockpile, or otherwise acquire or retain such agents or weapons designed to use such agents, there is no prohibition of possession of a production capability. Whilst possession of biological weapons would be a breach of the BTWC, the possession of dual-purpose weapons, equipment, or means of delivery would not. Nor are there provisions for intrusive verification and monitoring of compliance.

Over 115 nations are now party to the BTWC. An article of the convention required that, five years after the convention came into force, a conference of states parties should be held at Geneva, Switzerland, to review the operation of the convention, with a view to assuring that the purposes of the preamble and the provisions of the convention were being realized. Such reviews should take into account any new scientific and technological development relevant to the convention.

Such Review Conferences were held in 1980, 1986, and in September 1991. The First Review Conference, in 1980, occurred at the same time as information became available in the West about the outbreak of anthrax at Sverdlovsk in the Soviet Union discussed earlier in this chapter. The 1986 Second Review Conference took note of the advances in genetic engineering over the previous decade and agreed on four voluntary confidence-building measures:

1. declaration of all Category 4 laboratories and Ministry of Defence Category 3 laboratories;
2. declaration of unusual outbreaks of disease;
3. encouragement of publication of work;
4. encouragement of international conferences.

Although states parties are politically bound to make such returns annually, since 1987 only forty nations—about one-third of the signatories—have made returns. The quality of such returns was also variable, and the contribution to building international confidence slight. The geographic distribution of states parties to the BTWC shows that there is not complete participation in the BTWC in any part of the world—and this is equally true with regard to the annual confidence-building measure returns. The responses to the Second Review Conference on confidence-building (March, 1991) are shown below.

| | | |
|---|---|---|
| Argentina | France | Portugal |
| Australia | Greece | Qatar |
| Austria | Hungary | Senegal |
| Belgium | Ireland | Soviet Union |
| Bulgaria | Italy | (includes Ukraine, |
| Canada | Japan | Byelorussia) |
| Czechoslovakia | Korea | Spain |
| Chile | Mexico | Sweden |
| China | Mongolia | Switzerland |
| Denmark | Netherlands | Thailand |
| East Germany | New Zealand | Togo |
| Ecuador | Norway | United Kingdom |
| Finland | Poland | United States |
| | | West Germany |

Preparation for the Third Review Conference held in Geneva in September 1991 was extensive, with much activity both by governments and by non-governmental organizations such as SIPRI and the Federation of American Scientists, which meant that there was an extensive and intensive debate as to what measures the Third Review Conference should set out to achieve, in order to enhance the effectiveness of the BTWC.

### The Third Review Conference and Iraq

The Third Review Conference, held in Geneva, 9–27 September 1991, gained increased importance, occurring as it did in the year in which coalition forces faced the threat that Iraq might use chemical and biological weapons against them. It was assessed that Iraq had the capability to develop biological-warfare agents and appropriate countermeasures were taken. Subsequent to the 1991 Gulf War, the UN Security Council passed its Resolution 687, which requires:

1. Iraq to declare its stocks of nuclear, chemical, and biological weapons, missiles, agents, and production facilities;
2. the Iraqi nuclear, chemical, biological, and missile capabilities to be destroyed under the supervision of a UN Special Commission.

Iraq initially declared that it had no biological-warfare capability. However, following the first biological-weapons inspection of Iraq, which took place on 3–7 August 1991 at the Salman Pak facility, the UN Press Release of 14 August 1991 declared that:

1. Iraq had stated that research was undertaken on clostridium botulinum, clostridium perfringens, and bacillus anthracis.
2. Military research was later explained to comprise research which could be used for both defensive and offensive purposes.
3. Iraq had the capability to research, produce, test, and store biological-warfare agents.
4. Iraq admitted to having worked on anthrax and botulinum toxin.
5. Iraq handed over a collection of biological materials which could be developed as biological-warfare agents. This material included brucellosis and tularaemia.

The UN Special Commission has finalized its plan to ensure that the Iraqi capability to develop and produce biological weapons and biological weapons agents is eliminated. This compliance plan is detailed in S/22871/Rev. 1 dated 2 October 1991 and was approved in UN Security Council Resolution 715. This imposes a taut intrusive regime on Iraq which addresses *research* as well as development, production, and storage of biological weapons. The detailed requirements of this regime, with key elements in italics, and the inspection requirements are outlined in Table 5.6. The effectiveness of this plan will be extremely valuable in indicating to those concerned with improving the Review Conferences of the BTWC what sort of regime is necessary to be effective.

The Third Review Conference achieved several other advances:

1. a robust Final Declaration that reaffirmed the importance of the convention by declaring the signatories' 'continued determination for the sake of mankind, to exclude completely the possibility of the use of bacteriological (biological) agents and toxins as weapons, and their conviction that such use would be repugnant to the conscience of mankind';
2. an improved, well-focused, and extended confidence-building-measure regime outlined below;
3. the establishment of an *ad hoc* group of governmental

TABLE 5.6. *UN Compliance Monitoring and Verification Plan (UN SCR 715 and S/22871/Rev. 1)*

---

*Provisions relating to biological items*
- Declaration of any site or facility:
  - where *work with toxins or with micro-organisms* meeting WHO risk groups IV, III, or II is carried out;
  - where *work with genetic material coding* for toxins or genes derived from such micro-organisms is carried out;
  - having *a BL4 or BL3 containment laboratory*;
  - at which *fermentation* or other means for production of micro-organisms or toxins using vessels *larger than 10 litres individually or 40 litres in the aggregate* is carried out;
  - for the *bulk storage of toxins or of micro-organisms* in risk groups IV, III, or II;
- declaration of *any research, development, testing, or other* support or manufacturing *facility for specified equipment and other items*;
- declaration of *any imports, other acquisitions, or exports* of micro-organisms (risk groups IV, III, or II), toxins and vaccines as well as related equipment and facilities;
- provide a list of *all documents of a scientific and technical nature* published or prepared by any site or facility engaged in work *relating to toxins or micro-organisms* (risk groups IV, III, and II) including those of a theoretical nature;
- describe *all work on toxins or micro-organisms* (risk groups IV, III, and II) *as well as all work being conducted on the dissemination* of micro-organisms or toxins into the environment *or* on processes that would lead to such dissemination;
- provide information on *all cases of infectious diseases* affecting humans, animals or plants that deviate, or seem to deviate, from the normal pattern or are *caused by any micro-organism* (risk groups IV, III and II) *and* all such cases of similar occurrences caused *by toxins*.

---

*Iraq shall not*:
- *import items* without giving prior notice to the Special Commission;
- *conduct any activities in the field of micro-organisms or toxins except by civilian personnel* not in the employ of any military organization; any such activities shall be conducted openly;

TABLE 5.6. (*Continued*)

- *conduct activities on diseases other than those indigenous* to or immediately expected to break out in *its environment*;

- *conduct any breeding of vectors of human, animal, or plant diseases*;

- *possess at any one time more than one facility having a BL4 containment* laboratory;

- *possess at any one time more than two facilities having BL3 containment* laboratories.

*Equipment and other items relevant to biological and toxin warfare*
- *detection or assay systems* specific for risk groups IV, III, and II micro-organisms and toxins;

- *biohazard containment equipment*;

- *equipment for micro-encapsulation* of living micro-organisms;

- *complex media for the growth* of micro-organisms (risk groups IV, III, and II);

- *recombinant deoxyribonucleic acid (DNA), equipment and reagents* for its isolation, characterization, or production, and equipment and reagents for the construction of synthetic genes;

- *equipment for the release into the atmosphere of biological material*;

- *equipment for studying the aerobiological characteristics* of micro-organisms and toxins;

- *equipment for the breeding of vectors* of human, animal, or plant diseases.

The Special Commission shall have the right:
- to designate for inspection any site, facility, activity, material, or other item in Iraq;

- to carry out inspections, at any time and without hindrance, of any site, facility, activity, material, or other item in Iraq;

- to conduct unannounced inspections and inspections at short notice;

- to inspect any number of declared or designated sites or facilities simultaneously or sequentially;

- to designate for aerial overflight any area, location, site, or facility in Iraq.

TABLE 5.6. (*Continued*)

Iraq shall:

- accept unconditionally the inspection of any site, facility, activity, material, or other item declared by Iraq or designated by the Special Commission;
- accept unconditionally aerial overflight of any area, location, site, or facility designated by the Special Commission;
- provide immediate and unimpeded access to any site, facility, activity, material, or other item to be inspected;
- accept unconditionally the inspectors and all other personnel designated by the Special Commission and ensure their complete safety and freedom of movement;
- co-operate fully with the Special Commission and facilitate its inspections, overflights, and other activities under the Plan.

experts to identify and examine potential verification measures from a scientific and technical standpoint, this Group to meet in Geneva, 30 March–10 April 1992 and to aim to complete its work by the end of 1993;
4. the encouragement for states to implement measures to prevent international transfer of biological-warfare-related materials and technology (it was also declared that transfers relevant to the BTWC should only be authorized when the intended use is for purposes not prohibited under the convention; this provides a useful encouragement for the introduction of export-control measures).

The improved confidence-building-measure regime comprises:

1. exchange of data:
    (*a*) on research centres and laboratories;
    (*b*) on national biological defence research and development programmes;
2. exchange of information on outbreaks of infectious diseases and similar occurrences caused by toxins;
3. encouragement of publication of results and promotion of use of knowledge;

4. active promotion of contacts;
5. declaration of legislation, regulations, and other measures;
6. declaration of past activities in offensive and/or defensive biological research and development programmes since 1 January 1946;
7. declaration of vaccine-production facilities.

## VERIFICATION OF THE BIOLOGICAL AND TOXIN WEAPONS CONVENTION

The aim of any verification regime for the BTWC must be to build confidence that all states parties are compliant and, as far as possible, to enhance the deterrent effect of the verification regime so that a state party that is considering acquisition of a biological-warfare capability will judge that it will be found out and that the consequences of being found out are politically unacceptable. The agreement of the Third Review Conference of the BTWC in September 1991 to establish an *ad hoc* group to examine potential verification measures from a scientific and technical viewpoint is strongly supported. The *ad hoc* group is to aim to complete its work by the end of 1993. This presents a window of opportunity to strengthen significantly the BTWC. The *ad hoc* group is charged to evaluate proposed verification measures against the following criteria:

1. strengths and weaknesses based on the amount and quality of information provided and not provided;
2. ability to differentiate between prohibited and permitted activities;
3. ability to resolve ambiguities about compliance;
4. technology, material, manpower, and equipment requirements;
5. financial, legal, safety, and organizational implications;
6. impact on permitted activities and on proprietary commercial information.

The BTWC explicitly bans development of biological weapons but not research. It is suggested that this loophole be closed by bringing offensive biological-warfare research explicitly within the remit of the BTWC so that arguments cannot be made that a piece of work is offensive research not offensive development.

Although there would appear to be four potential models on which a verification regime for the BTWC might be based, three of them have particular deficiencies:

1. *The CWC verification regime model.* Much of the CWC verification regime would be inappropriate for the BTWC as biological weapons unlike chemical weapons, are already banned, so no state has yet admitted to possession of stocks of biological or toxin weapons, or to production facilities for such weapons. The potency of biological weapons is much greater than that of chemical weapons and, consequently, the quantities required for a minimum military capability are much smaller than for chemical weapons. This necessitates that a new regime is much more intrusive as the goal will be to find smaller quantities as well as storage, filling, and testing facilities.

2. *Verification of the BTWC confidence-building-measures model.* A possible verification regime would be one enabling the veracity of the confidence-building-measures declarations to be checked by visits to the facilities identified and to the countries making the declaration. Such a regime would increase the transparency of declared facilities, but it would be quite narrow in scope and make no provisions for visits to undeclared facilities.

3. *Investigation of allegations of use model.* The UN Secretary General, has, on occasion, sent in groups of experts to investigate allegations of use of chemical and biological weapons, most successfully in the Iran–Iraq War, when the Secretary General was invited by Iran to investigate alleged attacks by Iraq. Whilst the procedures to be used by the Secretary General's group of experts have been elaborated and are comprehensive, the provisions for ensuring rapid access to the suspect site have not advanced. The success or otherwise of such a regime would be fully dependent on the country in whose territory the suspect site is located inviting in the UN team of experts. Automaticity of access is an essential pre-requisite before this regime becomes worthy of further consideration as to whether extension to suspect sites other than sites of alleged use should be considered.

The fourth model appears to have more promise and is:

4. *UN Security Council Resolution 715 Iraq future compliance monitoring regime.* This regime is detailed in S/22871/Rev. 1 dated 2 October 1991 and is designed to give confidence that Iraq is in compliance with its obligations under UN Security Council Resolution 687 and 715.

Consequently, in so far as a verification regime is concerned, a sound basis on which to proceed would be the broad declaration and inspection regime developed by the UN Special Commission for the future compliance monitoring of Iraq. After all, there is no time limit to that future compliance monitoring and it is surely logical to argue that a regime devised to ensure that Iraq is compliant should be equally appropriate to ensuring that other states parties are compliant.

It is suggested, therefore, that the compliance monitoring regime detailed in S/22871/Rev. 1 of the 2 October 1991 should be taken as the starting-point for a compliance monitoring regime of the BTWC. Clearly, this regime should be modified in the light of experience from Iraq, as this should indicate which elements of the plan are the most effective as well as in which areas relaxation may be possible. It will also need to be modified according to the states party involved in respect of the permitted facilities. Nevertheless, the Special Commission compliance monitoring regime will serve as a useful starting-point for a non-punitive regime to be developed.

The central areas in any attempt to acquire an offensive biological-warfare capability should be identified and a verification regime targeted to maximize transparency in these areas, regardless of whether these are military or civil. Essential requirements for an offensive biological-warfare capability include the following:

1. the capability to produce and store quantities of biological and toxin warfare agents;
2. the ability to disseminate microbial organisms and toxins.

In addition to these two key capabilities, consideration should be given to what work is being carried out on micro-organisms that are not endemic to the area of concern. A difficulty immediately becomes apparent as one reason for studying micro-organisms and toxins that are not endemic to the country concerned is provision of defence against biological-warfare attacks. However, a more difficult problem is the discrimination between diseases and toxins that are endemic to the country concerned and those that are not. There are potential military advantages from utilizing an agent which occurs naturally as an attack is less likely to be recognized as such. Consequently, a division between micro-organisms that are endemic and those that are not does not contribute to confidence-building.

There is a need to achieve transparency on all work with dangerous pathogens and potent toxins. Even here, there are difficulties in that standards of containment are not internationally recognized and work that may in the United Kingdom or the United States be carried out in a high level of containment may well be carried out in other countries in little or no containment. There are, therefore, flaws and dangers in focusing exclusively on containments. There is potential benefit from a twin-track approach in which containment is one strand and work on a particular micro-organism or toxin is another.

A systems approach will be needed, as there is unlikely to be a single verification measure that in isolation will suffice to build confidence in compliance with the BTWC. A raft of measures such as those shown in Table 5.7 are likely to be necessary as confidence in compliance will be increased from internal consistency.

It will be important to have an inspection regime complementing the broad declarations required. This should certainly cover declared facilities, but the way in which it is to cover undeclared facilities will doubtless be the subject of considerable debate. The regime for inspections of undeclared sites in Iraq will be too demanding both financially and in terms of confidentiality of legitimate activities, but considerable intrusion will be needed to deter violation. A possible answer could be a combination of inspections of declared sites, and end-user verification of export bans for potentially biological-warfare-related materials transferred to undeclared sites.

## WEB OF DETERRENCE

It is becoming evident that the deterrence of states from acquiring biological weapons requires the creation and maintenance of a web of deterrence comprising three key elements:

TABLE 5.7. *Potential BTWC verification measures: A systems approach*

| Activity | Potential verification measures[a] | | | | | | |
|---|---|---|---|---|---|---|---|
| | Declared programmes | Containment levels[b] | Declared materials[c] | Declared fermenter sizes | Safety/medical measures[d] | Analytical tests | Audit trail |
| Research and development facilities | + | + | + | + | + | + | + |
| Production facilities | + | + | + | + | + | + | + |
| Test chambers | + | + | + | − | + | + | + |
| Testing grounds | + | + | + | − | + | + | + |

[a] Aim is to check internal consistency and therefore build confidence.
[b] National regulations needed, showing required containment levels.
[c] Materials related to a list of putative biological and toxin weapons agents.
[d] Measures related to assessments for control of substances hazardous to health, and ACDP.

1. the provision and maintenance of effective biological-warfare defence protective measures, thereby reducing the military advantage to be gained from use of biological weapons;
2. the improvement of biological-weapons arms controls through universal compliance with improved confidence-building measures and, if an effective regime can be devised, intrusive inspection and monitoring regimes that lead a potential acquirer of biological weapons to judge that his acquisition of biological warfare will be detected; automaticity of immediate access to investigate allegations of non-compliance and of use is also required;
3. the imposition and improvement of export controls and monitoring for biological-warfare agents and technology.

## THE WAY AHEAD

The 1980s saw an increasing proliferation in the numbers of states that either possess or have an interest in acquiring a chemical- or biological-warfare capability. The number of nations considered to have or to be acquiring a chemical-warfare capability is reported as being in excess of twenty,[19] and the Defence White Paper in July 1991 stated that ten nations have or are developing a biological-warfare capability[20] and are undeterred by the BTWC's prohibition of biological warfare. When the potential chemical- and biological-warfare spectrum (Fig. 5.1) and increase in potency of the biological-warfare end of the spectrum (Fig. 5.3) are considered, it is evident that states which acquire a chemical-warfare capability are likely to be attracted by a biological-warfare capability, because of the smaller quantities required, the ease with which a smaller facility can be hidden, and the difficulty of distinguishing biological-warfare attacks involving agents that produce endemic diseases from natural outbreaks of such diseases. It is therefore important to strengthen the provisions of the BTWC and impose effective export controls to deter biological-warfare proliferation.

[19] US Information Service, 16 July 1990; *Jane's Defence Weekly*, 14 July 1990, p. 51.
[20] Statement on the Defence Estimates, *Britain's Defence for the 90s*, Cmd 1559 (London: HMSO, July 1991).

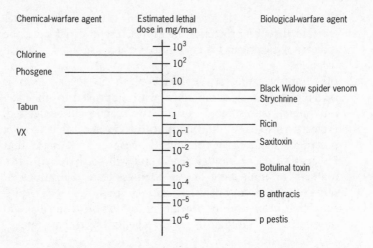

FIG. 5.3. Toxicity of chemical- and biological-warfare agents
(From SIPRI report, 1970.)

The threat that Iraq might attack the coalition forces using biological weapons and the actions taken by the coalition forces to counter that eventuality have greatly heightened the awareness around the world of the danger and—unfortunately—the potential utility of biological warfare. Fortunately, biological warfare was not used by Iraq and the subsequent activities of the UN Special Commission to eliminate Iraq's capability to produce biological weapons offers the exciting prospect that the world will recognize the dangers posed by biological warfare and will agree measures to improve the BTWC, taking into account the experience gained by the UN Special Commission. This is a unique opportunity, and we should all do what we can to ensure that it is grasped. If we fail to do so, it is possible that biological warfare will be used in a future conflict by a biological-weapons possessor who may judge that the benefits outweigh the repercussions of getting caught. International will is needed to ensure that this whole class of warfare is effectively banned in a way in which a potential biological-weapons possessor would judge that his cheating would be detected with consequential unacceptable international repercussions to the cheater.

The way forward then is undoubtedly to create an effective web of deterrence by:

1. improving biological-warfare protective measures and medical countermeasures, so minimizing the utility of biological warfare to a potential aggressor;
2. improving biological-warfare arms-control regimes so that a potential acquirer will judge that his activities will be detected and the consequential international repercussions will be unacceptable;
3. improving export controls for biological-warfare agents and technology so that acquisition of a biological-warfare capability is made more difficult.

# 6

# Nuclear Proliferation in the Middle East: The Next Chapter Begins

## LEONARD S. SPECTOR

DURING 1991 a new chapter in the nuclear history of the Middle East began to unfold, as the locus of regional proliferation concerns shifted dramatically. On the one hand, Iraq's massive nuclear-weapons programme of the 1980s was exposed and brought to a halt as a result of the 1991 Gulf War and its aftermath. At the same time, however, Iran's bid for nuclear arms accelerated dramatically; Syria, for the first time, demonstrated an interest in obtaining nuclear weapons; and Algeria's efforts to establish a nuclear infrastructure with military overtones were discovered. Moreover, a new estimate of Israel's capabilities suggested that its nuclear arsenal was far larger than previously assumed.[1]

As these trends evolved, the NPT was shown to have been ineffective in restraining Iraq's quest for nuclear arms, raising questions about the utility of the pact as a curb on proliferation elsewhere in the region. Bahrain, Egypt, Iran, Iraq, Jordan, Kuwait, Libya, Saudi Arabia, Syria, and Tunisia are parties to the accord, although neither Israel nor Algeria has joined it. By the end of 1991, efforts were under way to strengthen the treaty significantly. This offered the prospect that, whatever its past failings, the newly fortified pact might yet prove capable of checking the programmes of the next wave of Middle Eastern nuclear aspirants.[2]

This chapter will review these important new developments.

---

[1] Despite Muammar Khadafi's repeated expressions of interest in acquiring nuclear weapons, Libya's pursuit of this goal appears to have been stymied since the late 1980s, and Libya's nuclear programme will not be analysed further here.

[2] International attention has focused recently on President Bush's June 1991 Middle East Arms Control Initiative and the proposal by Egyptian President Hosni

## IRAQ

During the spring and summer of 1991, inspections mandated by the UN Security Council following Iraq's defeat in the 1991 Gulf War revealed that Iraq had been pursuing a multi-billion-dollar covert nuclear-weapons programme for nearly a decade. This programme, codenamed 'Petrochemical 3', was in direct violation of Iraq's obligations under the NPT, which it ratified in 1969, and on 18 July and 20 September 1991 the IAEA formally condemned Iraq for infringing the pact.

The inspections, conducted by the IAEA under the auspices of the UN Special Commission on Iraq, were authorized in April 1991 by Security Council Resolution 687. The resolution established procedures for the destruction of Iraq's non-conventional-weapons and ballistic-missile capabilities and called for a subsequent monitoring programme to prevent Iraq from rebuilding them.

Among the startling discoveries unearthed by the UN–IAEA inspections were that Iraq had simultaneously pursued four different technologies for enriching uranium to weapons grade, and had advanced furthest with the electromagnetic-isotope-separation (EMIS) process.[3] Neither the United States, nor any other foreign government, had detected the EMIS programme, in part because it did not rely heavily on imported high-technology equipment, whose acquisition might have given the effort away.[4] Indeed,

Mubarak for a weapons-of-mass-destruction-free zone in the region. Whatever the merits of these proposals, few would dispute that their implementation is likely to be many years distant, coming, if ever, only after substantial progress has been made towards a regional peace settlement—and perhaps only after a generation of peace.

NPT inspections were strengthened in late 1991. Though this will have little impact on Israel's nuclear capabilities, its utility as a non-proliferation measure elsewhere in the region could be substantial. For this reason, the NPT is the focus of discussion here.

[3] To produce weapons-grade uranium, natural uranium must be 'enriched' (i.e. it must be processed to increase the concentration of easily fissioned uranium-235 atoms, from their naturally occurring concentration of 0.7% to 80% or more). The other enrichment methods Iraq pursued were the gas centrifuge, gaseous diffusion, and chemical-separation processes.

[4] By 1988 US intelligence analysts were aware that Iraq was pursuing the centrifuge process, because they had detected Iraq's efforts to obtain high-technology items from Western Europe needed to support the effort. US analysts had learnt of other Iraqi activities apparently associated with uranium enrichment, but mistakenly thought that these were linked to the centrifuge programme. Thus the US specialists did not pursue the possibility that Baghdad was engaged in an

the EMIS programme might have remained hidden from the UN–IAEA inspection teams but for the fact that it was revealed by an Iraqi nuclear engineer who had defected to US forces after the war.

As a result of their inspections through the early autumn of 1991, the UN–IAEA inspectors concluded that Iraq had built and operated only a small number of EMIS units (known as 'calutrons') at an experimental facility at the Tuwaitha Nuclear Research Centre near Baghdad and at a partially completed industrial-scale facility in Tarmiyah. It appeared that Iraq had produced only a small quantity of non-weapons-grade enriched uranium using the process;[5] had it been able to complete the Tarmiyah facility, it might have been able to produce enough weapons-grade uranium for its first nuclear device in as little as two years. The inspectors also determined that a replica of the Tarmiyah plant was under construction at Ash-Sharqat, indicating that Iraq had eventually hoped to produce enough weapons-grade material for two or three nuclear devices per year using the EMIS method.

The inspections also exposed a series of previously unknown facilities for producing and processing uranium into a number of uranium compounds used in the EMIS and other enrichment processes. In addition, the inspectors discovered evidence that Iraq had secretly extracted a small amount of plutonium from uranium irradiated in the Soviet-supplied IRT-5000 reactor—a facility that was under IAEA safeguards—without informing the agency, as required under the Iraq–IAEA safeguards agreement.[6]

enrichment effort involving multiple processes. They also discounted the possibility that Iraq might employ the EMIS method, since they considered it obsolete and therefore unattractive to a would-be nuclear power. (Based on discussions with US officials, summer 1991.)

[5] In July 1991 some US officials were quoted as stating that Iraq had produced enough weapons-grade material using the EMIS method for one or two nuclear devices, an allegation originally made by the defector who disclosed the programme (see E. Sciolino, 'Word of Iraqi Nuclear Effort is a Mixed Blessing for Bush', *New York Times*, 10 July 1991; R. J. Smith, 'Reassessing Iraqi Nuclear Capability', *Washington Post*, 10 July 1991). These statements apparently proved to be inaccurate, since the material was never located and the UN–IAEA inspectors, after reviewing the status of Iraq's EMIS programme, concluded that it was not sufficiently advanced when the 1991 Gulf War began to have produced enough weapons-grade material for a nuclear device. Iraq claimed to have produced only 4 kg of enriched product, an amount that included only gram quantities of uranium enriched to more than 20%.

[6] M. Hibbs, 'Loophole in Safeguards Rules Allowed Iraq to Reprocess', *Nucleonics Week*, 15 Aug. 1991, p. 1.

In addition, the inspections led to the seizure of thousands of pages of Iraqi documents related to its nuclear activities. The documents revealed details of Baghdad's efforts to design a nuclear explosive device and to test its non-nuclear components, thereby providing unambiguous evidence of Iraq's ultimate objective and belying the repeated claims of Iraqi officials that the country's nuclear programme was intended for non-military purposes. The documents also described Baghdad's extensive efforts to obtain technology abroad for various facets of its nuclear programme.

The seized documents further revealed that Iraq had planned to produce large quantities of lithium-6, a material used exclusively for the production of 'boosted' atomic bombs and thermonuclear, or 'hydrogen' bombs. In addition, the inspectors found that Iraq had conducted tests aimed at arming a ballistic missile with a nuclear warhead. These two initiatives—as well as the multi-pronged enrichment programme—demonstrated that Iraq had hoped to become a full-fledged nuclear-weapon state as rapidly as possible.

Iraq refused, however, to tender all relevant documents and, in one confrontation in August, regained custody of a significant quantity of documents that had been seized by the UN–IAEA team. It later returned only a portion of the documents involved. The inspectors also found repeated instances in which documents had been removed from inspected premises to prevent scrutiny by the UN–IAEA teams.

Irrespective of the outcome of this matter, the inspectors' discoveries provide important new insights into the history of the Iraqi nuclear programme. It had been widely assumed that Iraq had done little to pursue its nuclear ambitions during the period between Israel's destruction of the Osiraq reactor outside Baghdad in June 1981 and the latter stages of the Iran–Iraq War—despite Iraqi President Saddam Hussein's call shortly after the Israeli raid for the international community to 'assist the Arabs in one way or another to obtain the nuclear bomb in order to confront Israel's existing bombs'.[7]

---

[7] 'Iraq Asserts Arabs Must Acquire Atom Arms as a Balance to Israel', *New York Times*, 24 June 1989. One of the only indications of Iraqi efforts to pursue this goal in the mid-1980s was an episode described in a 1984 Italian prosecution, in which a number of senior Iraqi military figures apparently expressed interest in obtaining a quantity of plutonium sufficient for several nuclear weapons from an

The UN–IAEA inspectors determined, however, that, in fact, Iraq restarted its nuclear-weapons programme in 1982, launching an effort to master uranium enrichment using the EMIS method and a second alternative based on high-speed gas centrifuges. During the mid-1980s, as this work and other laboratory-scale work continued, construction began on a major uranium-processing plant, that was completed in 1989. In the late 1980s the pace of the programme accelerated, as Iraq experimented with two other, subsequently abandoned, enrichment methods; attempted to build industrial-scale facilities at Al-Jezera and other locations to produce feedstock for the various enrichment processes; and worked to develop the design of a nuclear bomb and test its non-nuclear components. In February 1990 the first stages of a full-scale EMIS enrichment plant were inaugurated at Tarmiyah; a new facility was opened at Al-Atheer, consolidating nuclear design work previously performed at several other installations; and construction began on a facility for the serial production of uranium-enrichment centrifuges at Al-Furat.

Under the NPT, Iraq was required to declare all of its nuclear installations and materials (except uranium ore and ore concentrate) and submit them to IAEA inspection to allow verification that they were not being used for the development of nuclear weapons. As noted earlier, Iraq repeatedly violated this obligation by failing to disclose, for example, the construction and operation of the EMIS units at Tuwaitha and Tarmiyah; the production of uranium feedstocks for the enrichment process; and the production of some hundred tons of uranium dioxide, a partially processed uranium product, at the Al-Jezera plant. Moreover, even at the IRT-5000 reactor and nearby laboratories, which were under IAEA monitoring, Iraq was able to circumvent the agency's controls and covertly produce a small quantity of plutonium.

The failure of the IAEA to detect any of these violations of its rules has raised serious questions about the effectiveness of the

Italian arms-smuggling ring purporting to have such material for sale. The deal—which was certainly a hoax—fell through when the smugglers were unable to produce samples of the nuclear material (see L. S. Spector, *The New Nuclear Nations* (New York: Vintage, 1985), 44–54). In August 1990 evidence began to emerge that Iraq's smuggling efforts to support its centrifuge-enrichment programme were well under way by 1987, suggesting a mid-1980s start-up date for that programme (see M. Hibbs, 'Trail of Iraqi Nuclear Commerce Leads to Two Swiss Companies', *Nucleonics Week*, 23 Aug. 1990, p. 13).

IAEA system and the NPT. (These are discussed in the section about the NPT, below.)

The revelations about Iraq's clandestine nuclear activities also raise serious questions about the adequacy of nuclear intelligence-gathering by the United States and other concerned nations. The failure to detect the Iraqi electromagnetic enrichment programme—and the similar failure, discussed later, to detect for several years Algeria's construction of a sizeable research reactor—inevitably raise doubts about assessments of nuclear programmes in other countries which, like Iraq, may be prepared to skirt their non-proliferation commitments.

The majority of the installations involved in Iraq's nuclear-weapons programme were destroyed by US bombing raids during the conflict. Other facilities—many of which had been unknown to the United States and its coalition partners—were dismantled by Iraq, itself, after the war, in an effort to deceive the UN–IAEA inspectors about the scope of the country's nuclear activities. Many pieces of specialized equipment were also destroyed by Iraq or the UN–IAEA teams. In addition, weapons-grade uranium that Iraq had obtained from France and the Soviet Union during the 1980s was placed in IAEA custody for removal from the country.

None the less, presumably at the order of President Saddam Hussein, Iraqi officials steadfastly resisted the UN effort to dismantle the country's nuclear programme, repeatedly providing false and incomplete data in declarations required by Resolution 687 and, whenever possible, hiding nuclear material to prevent its examination and destruction by the inspection teams. Although pressure from the Security Council, the United States, and several other major powers ultimately led Iraq to comply with many UN demands for information and for access to facilities, it appeared that Iraq had been able to withhold some of its nuclear equipment and material from the UN–IAEA inspection teams, along with the bulk of the programme's technical documents. The UN–IAEA inspectors were particularly concerned that they had not unearthed all of the equipment Iraq was thought to have obtained for its nascent programme to enrich uranium through the use of high-speed-gas centrifuges.

Given the continued availability of the Iraqi scientists and technicians and of much of the documentary record of the country's nuclear programme, including research results and engineering drawings, the potential clearly remains for Iraq to reconstitute its nuclear programme at some future time. To prevent a resurgence of Iraq's nuclear potential, Security Council Resolution 687 provided for the continued monitoring of Iraq's capabilities for developing and manufacturing weapons of mass destruction and ballistic missiles. On 11 October 1991 the Security Council adopted Resolution 715, setting out the details of this monitoring programme. The resolution, among other steps, prohibits Iraq from engaging in all nuclear activities except those associated with the use of nuclear materials for medical purposes.

If the Security Council remains steadfast in its application of these monitoring procedures, it appears likely to succeed in arresting the Iraqi nuclear-weapons effort. This would make Iraq's acquisition of nuclear arms before the turn of the century most unlikely. On the other hand, it must be recognized that, as long as Saddam Hussein or a similarly militaristic leader remains in power in Baghdad, efforts to outwit the UN monitoring effort and secretly rebuild the Iraqi nuclear programme will continue. So far, the United Nations has won each round in the continuing battle of wills with Baghdad, forcing it to disclose more and more of its secrets and to accept increasingly intrusive monitoring.

In several key confrontations during the summer of 1991, it was necessary for the United States to back up the Security Council's demands on Iraq with the threat of force. How the November 1992 US presidential election will affect Washington's resolve remains to be seen. If the United States eases its pressure, the effectiveness of the UN monitoring effort could wane, opening the way to a resurgence of Iraq's bid for nuclear arms.

## IRAN

Like Iraq, Iran is a party to the NPT, but it, too, is apparently pursuing a secret nuclear-weapons programme.

Although unable to complete the German-supplied nuclear-power reactors at Bushehr, whose construction began under Shah Mohammed Reza Pahlavi, the Iranian Revolutionary Government has maintained the nuclear-research base that it inherited from him.[8] Work at the Tehran Research Centre, for example,

---

[8] The partially completed reactors were severely damaged by Iraqi bombing raids during the Iran–Iraq War, and Germany, concerned over Iran's possible interest in nuclear weapons, has refused to assist in finishing the plants. Iran has been said to be seeking help for the project from China.

apparently continued without major interruption after Ayatollah Khomeini took power in 1979, permitting the training of specialists and the use of a small US-supplied research reactor at the centre, which remains under IAEA Agency safeguards.[9] Presumably, specialists at the centre had access to the investigations undertaken during the Shah's reign, including the fruits of what appears to have been a modest undeclared nuclear-weapons research effort.

Further demonstrating the Khomeini regime's commitment to continued nuclear research, in 1984—in the midst of the Iran–Iraq War—it opened a new research centre at Isfahan, for which ground had been broken under the Shah.[10] In 1987, moreover, it signed a $5.5-million contract with Argentina, under which the latter agreed to supply new, non-weapons-grade, 20-per-cent-enriched uranium fuel for the Tehran research reactor.[11] Tehran is also reported to have acquired substantial quantities of uranium concentrate, or 'yellowcake', from South Africa in 1988–9. The material might have been acquired for enrichment at some future date by Iran, or possibly for enrichment in Pakistan and for subsequent use there or in Iran.[12] This would be consistent with past indications of possible nuclear ties between the two neighbouring Islamic countries.

By the late 1980s, indications that Iran was pursuing a nuclear-weapons programme began to mount. In a February 1987 speech at the Atomic Energy Organization of Iran, Ali Khamenei, then Iran's president, reportedly declared:

Regarding atomic energy, we need it now. . . . Our nation has always been threatened from outside. The least we can do to face this danger is to let our enemies know that we can defend ourselves. Therefore, every step

[9] See A. Etemad, 'Iran', in H. Mueller, *European Non-Proliferation Policy* (Oxford: Oxford University Press, 1987). Etemad was the chairman of the Atomic Energy Organization of Iran during its period of greatest activity under the Shah.
[10] Ibid. 9; 'Iranian Reactor to Go Critical', *Nuclear Engineering International* (Dec. 1984).
[11] The 'reactor' referred to is a subcritical training unit (i.e. one that will not sustain a nuclear chain reaction.) See 'Iran to Receive Nuclear Technology, Know-How', Noticias Argentinas, 18.00 GMT, 18 May 1987, trans. in *Foreign Broadcast Information Service (FBIS)/Latin America*, 19 May 1987; 'Argentina Confirms Deal for Work on Bushehr', *Nuclear News* (July 1987), 54. See R. Kessler, 'Argentina to Enforce Curbs on Nuclear Trade with Iran', *Nucleonics Week*, 14 Mar. 1987, p. 12.
[12] M. Hibbs, 'Bonn will Decline Teheran Bid to Resuscitate Bushehr Project', *Nucleonics Week*, 2 May 1991, p. 17.

you take here is in defence of your country and your revolution. With this in mind, you should work hard and at great speed.[13]

In October 1988, shortly after the ceasefire in the war with Iraq, Ali Akbar Hashemi-Rafsanjani, then the speaker of the Iranian parliament and commander-in-chief of Iran's armed forces, was more explicit. In an address to a group of Iranian soldiers, he declared:

With regard to chemical, bacteriological, and radiological weapons training, it was made very clear during the war that these weapons are very decisive. It was also made clear that the moral teachings of the world are not very effective when war reaches a serious stage and the world does not respect its own resolutions and closes its eyes to the violations and all the aggressions which are committed in the battle field.

*We should fully equip ourselves both in the offensive and defensive use of chemical, bacteriological, and radiological weapons.* From now on you should make use of the opportunity and perform this task.[14]

Confirming these expressions of interest in nuclear arms, in early 1989, Rear-Admiral Thomas E. Brooks, Director of Naval Intelligence, declared in congressional testimony that Iran was 'actively pursuing' a nuclear-weapons capability. He provided no details in his public testimony, however.[15]

It appears that Iran's efforts to develop a nuclear-weapons capability accelerated significantly in 1991. According to US officials, Iran has turned to Western Europe for the necessary hardware and technology, employing clandestine purchasing

[13] D. Segal, 'Atomic Ayatollahs', *Washington Post*, 12 Apr. 1987. Segal took the quoted material from *Nameh Mardom*, a Tudeh party newspaper published in Stockholm, Sweden, which claims to have obtained the text from a tape-recording made by one of those present at Khamenei's address and subsequently smuggled out of the country.

[14] 'Hashemi-Rafsanjani Speaks on the Future of the IRGC [Iranian Revolutionary Guards Corps]', Tehran Domestic Service, 09.35 GMT, 6 Oct. 1988, trans. in *FBIS-NES*, 7 Oct. 1988, p. 52 (emphasis added). For evidence that some officials of the Iranian Revolutionary Government were interested in nuclear arms as early as 1979, see D. Segal, 'Iran Speeds Up Nuclear Bomb Development', *Journal of Defense and Diplomacy*, 6/6 (1988), 52; Segal, 'Atomic Ayatollahs'. See also L. S. Spector, *Going Nuclear* (Cambridge, Mass.: Ballinger Publishing Company, 1987), 56.

[15] Testimony of Rear-Admiral Thomas A. Brooks, Director of Naval Intelligence, before the Subcommittee on Seapower, Strategic, and Critical Materials of the Committee on Armed Services, US House of Representatives, 22 Feb. 1989 (mimeo).

networks similar to those used by Iraq and Pakistan.[16] US officials also assume that Iran has begun to conduct research on the production of weapons-grade nuclear materials; such activities, US officials believe, have little relation to a peaceful nuclear programme and are seen as a clear sign of a nuclear-weapons effort.[17] Such research is said by some reports to be centred in Qazvin, near the Caspian Sea, and to be under the control of the Revolutionary Guards, rather than the Atomic Energy Organization of Iran, which is responsible for other aspects of the country's nuclear programme.[18] Iran is also apparently receiving extensive nuclear assistance from China. Both countries claim the aid is exclusively for peaceful purposes, but, even if this is true, it could bolster Iran's nuclear-technology base and indirectly support its weapons programme, a concern that has triggered US objections to the Chinese assistance.[19] Reportedly Tehran is training nuclear technicians in China under a recent agreement for co-operation, which may also include assistance on reactor construction at Iran's Isfahan nuclear-research centre.[20]

It has also been learnt that China supplied Iran with a small-scale calutron—the type of equipment used in Iraq's EMIS-enrichment programme for the improvement of uranium to weapons grade. Apparently, the unit that China sold to Iran is not

[16] Personal interviews.

[17] J. Mann, 'Iran's Nuclear Plans Worry US Officials', *Los Angeles Times*, 27 Jan. 1991; R. J. Smith, 'Officials Say Iran is Seeking Nuclear Weapons Capability', *Washington Post*, 30 Oct. 1991. In October 1991 an unnamed US aide was quoted as stating, 'Obviously we are very concerned and we think, in fact, that Iran is trying to develop a nuclear weapons program' (see C. Aldinger, 'Iran not Close to Developing Nuclear Arms', *Associated Press*, 31 Oct. 1991).

[18] R. Evans and R. Novak, 'Beijing's Tehran Connection', *Washington Post*, 26 June 1991; 'The China–Iran Nuclear Cloud', *Mednews*, 22 July 1991 (giving the precise location of Iran's nuclear-weapons research centre as Moallem Kalayeh, in the Elburz Mountains, just north of Qazvin).

[19] 'Iran Confirms Nuclear Cooperation with China', *United Press International*, 6 Nov. 1991; R. J. Smith, 'China–Iran Nuclear Tie Long Known', *Washington Post*, 31 Oct. 1991.

[20] Ibid.; Smith, 'Officials Say Iran is Seeking a Nuclear Arms Capability'; Testimony of Gary Milhollin, Director Wisconsin Project on Arms Control, Hearings on US–China Relations, US Senate Foreign-Relations Committee, 13 June 1991. The Isfahan Centre houses a zero-power reactor, known as a 'subcritical' assembly, which is used for training. The agreement was apparently signed in June 1990. See Smith, 'Officials Say Iran is Seeking a Nuclear Arms Capability'; but see Evans and Novak, 'Beijing's Tehran Connection', indicating the agreement was signed in 1985.

powerful enough to process uranium, but is a smaller model for separating medical isotopes. None the less, Iran will undoubtedly glean practical information about the EMIS process from the Chinese equipment. Summarizing US concerns, Assistant Secretary of State Richard H. Solomon stated in a congressional hearing, 'While the Chinese may not be selling finished weapons, they may be transferring certain technologies or information ... [which] is unacceptable to us.'[21]

Iran is also seeking to purchase a research reactor from India.[22] The unit is said to have a power level of only 10 megawatts, which—unless the facility were upgraded—would not be sufficient for the production of substantial quantities of plutonium. The unit will also be subject to IAEA inspection, which would make it difficult for Iran to use the reactor to produce weapons-relevant quantities of plutonium surreptitiously. None the less, the reactor could substantially bolster the country's nuclear-research base at a time when Iran is believed to be actively pursuing a nuclear-weapons capability.

Amidst these developments, Iran's Deputy President, Ayatollah Mohajerani, provided further evidence of his country's interest in acquiring nuclear arms, declaring in an October 1991 interview distributed by the official Iranian news agency that, 'because the enemy has nuclear facilities, the Muslim states too should be equipped with the same capacity'.[23]

Iran's nuclear-weapons programme appears to be still in its infancy, suggesting that it will be a number of years—and possibly a decade—before the country may realize its nuclear aspirations.[24] None the less, after Iraq's unexpected advances, predictions about the pace of Iran's future nuclear achievements must be made with caution.

SYRIA

Syria's nuclear programme dates back to 1977, when the Syrian Atomic Energy Commission was formed. According to a former

[21] Smith, 'China–Iran Nuclear Tie Long Known'.
[22] S. Coll, 'Iran Reported Trying to Buy Indian Reactor', *Washington Post*, 15 Nov. 1991.
[23] Smith, 'Officials Say Iran is Seeking Nuclear Weapons Capability'.
[24] Aldinger, 'Iran not Close to Developing Nuclear Arms'.

aide to Syrian President Hafez al-Asad, it began training specialists abroad in the late 1970s, particularly in France. In 1981, he stated, Damascus signed a nuclear-co-operation agreement with India. Little has been revealed, however, about India's possible sharing of nuclear technology with its Syrian partner.

In the mid-1980s Syrian Foreign Minister Mustafa Tlas declared that the Soviet Union had 'guaranteed' it would give Damascus nuclear weapons if Israel employed nuclear arms against Syria.[25] Whatever the truth to the assertion, it represented one of the few occasions in recent decades when Syria has suggested an interest in acquiring nuclear arms.

Syria is not known to have any significant nuclear facilities. In the autumn of 1991, however, new information indicated that Syria was seeking to build a nuclear infrastructure with military objectives in mind. At a September 1991 meeting at the National Academy of Sciences, in Washington, DC, Bradley Gordon, Assistant Director of the US Arms Control and Disarmament Agency, listed Syria among countries having 'nuclear programs with suspicious intentions'. Other US officials indicated in October 1991 that Syria was seeking to purchase a research reactor from China, and, in late November, China announced that it was planning to sell such a facility to the IAEA for subsequent transfer to Syria.[26]

China stated that the reactor would have a power level of 30 kilowatts. Such a reactor would be useful only for training purposes and could not produce material for nuclear arms. None the less, the unit represents Syria's effort to acquire a nuclear installation, and Gordon's comments suggest that it may be the first step on the road towards nuclear arms, however long that road may be. Syria is obligated to submit the reactor and any other nuclear facilities and materials (other than uranium ore and ore concentrate) to IAEA inspection, because it is a party to the NPT. This should further reduce the proliferation threat posed by the facility, although Israel is likely to be concerned by the very fact that Syria has begun to develop a nuclear infrastructure.

[25] 'War of Liberation', *New York Review of Books*, 22 Nov. 1984, p. 36; N. Roland, 'Soviets Reportedly Offer Nuclear Help to Syria', *United Press International*, 28 Nov. 1985. See also Z. Schiff, 'Dealing with Syria', *Foreign Policy* (summer 1984), 94.

[26] *Associated Press*, 28 Nov. 1991.

As for Syria's possible motivation for developing nuclear arms, despite its recent gains in Lebanon and the decimation of its principal Arab antagonist in the 1991 Gulf War, in the longer term Syria faces increasing isolation, given the withdrawal of support by its superpower ally, the Soviet Union. As has been true for other radical regimes around the globe in similar straits, such circumstances have apparently enhanced the attractiveness of acquiring a nuclear option. Though there are no additional details concerning Syria's nuclear plans, the fact that it has now been identified by the United States as apparently having an undeclared interest in acquiring such weapons is, in itself, a development of considerable importance.

### ALGERIA

Throughout the 1980s Algeria's nuclear programme[27] was widely assumed to be limited to uranium-exploration activities and the operation of a nuclear-research centre on the outskirts of Algiers, where a small (1-megawatt) research reactor supplied by Argentina was commissioned in March 1989.[28] Although Algeria had not signed the NPT, that reactor was placed under IAEA safeguards, and there was little to suggest any Algerian interest in acquiring nuclear arms.

In January 1991, however, US intelligence agencies obtained evidence that Algeria was secretly building a sizeable research reactor near the town of Ain Oussera, some 155 miles south of Algiers.[29] The partially completed facility, which was said to

[27] This section is based in part on a chapter from L. S. Spector and J. R. Smith, *Nuclear Threshold* (Boulder, Colo.: Westview Press, 1992).

[28] The research centre, known as the Centre des Sciences et de la Technologie Nucléaire, was established in the mid-1960s, as the successor to the Institut d'Études Nucléaires, founded in 1958, when Algeria was still a French colony. For a discussion of the early years of the Algerian nuclear programme, see A. Mustafa, 'Nuclear Fuel Resources in the Arab World', a paper presented at the Second Arab Energy Conference, 6 Mar. 1982, Doha, Qatar, quoted in T. W. Graham, M. M. Miller, D. M. Poneman, G. W. Rathjens, J. M. Star, and R. L. Williamson, jr., 'Nuclear Power and Nuclear Proliferation in the Islamic Middle East', A Study Prepared for the United States Congress Office of Technology Assessment, 3 June 1983 (unpublished).

[29] Some US State Department officials apparently learnt of the contract for the sale of the reactor in late 1988, but did not reveal this information to other US government agencies (see E. Sciolino, 'Algerian Reactor: A Chinese Export', *New York Times*, 15 Nov. 1991).

be ringed with anti-aircraft missiles, apparently had been under construction for several years and was being built with Chinese assistance.[30]

Based on analysis of reconnaissance satellite photos of the unit's cooling system, some US estimates of the reactor's size indicated that it was 40 megawatts or larger—a size more appropriate for plutonium production than for peaceful research and training. The secrecy surrounding the project intensified suspicions as to its ultimate purpose and strongly suggested that the reactor was not going to be placed under IAEA safeguards; since neither Algeria nor China was a party to the NPT, neither country had a legal obligation to take this step. While China had repeatedly pledged since 1984 that all of its nuclear exports would be placed under IAEA monitoring, it was feared that Beijing might renege on these commitments and quietly help Algeria to launch a nuclear-weapons programme.[31]

None the less, conclusions concerning the threat posed by the reactor remained tentative. US analysts did not observe any other indications of an Algerian nuclear-weapons effort, and some believed the satellite photos showed the Ain Oussera reactor to be of a smaller, less suspect size. Moreover, there appeared to be no obvious rationale for an Algerian nuclear-weapons programme, since the country did not face an external threat that might justify such a drastic response.

[30] Construction reportedly began at the end of 1986 (see 'Sources Say Reactor Data to be Released', *Al-Sharq Al-Awsat*, 1 May 1991, trans. in *JPRS-TND*, 31 May 1991, p. 21).
[31] B. Gertz, 'China Helps Algeria Develop Nuclear Weapons', *Washington Times*, 11 Apr. 1991; J. Mann, 'China may be Giving A-Arms Aid to Algeria', *Los Angeles Times*, 12 Apr. 1991; R. J. Smith, 'China Aid on Algerian Reactor', *Washington Post*, 20 Apr. 1991; M. Hibbs, 'Cooling Towers are Key to Claim Algeria is Building Bomb Reactor', *Nucleonics Week*, 18 Apr. 1991, p. 7; personal interviews with US officials. In contrast to the secrecy surrounding the Chinese-supplied reactor, Argentina had promptly announced the sale of the 1-megawatt 'NUR' reactor at the Centre des Sciences et de la Technologie Nucléaire. The announcement was made in 1985, when the agreement for the transfer of the unit to Algeria was signed. Argentina also sold Algeria enriched uranium fuel for the NUR reactor. However, given the delays Argentina experienced in expanding its indigenous uranium-enrichment plant at Pilcaniyeu, it is unlikely that Argentina had produced sufficient enriched material to fuel the NUR unit by the time it was commissioned in 1989. Thus it is possible that the fuel Argentina transferred for the facility was, in fact, Chinese-origin enriched uranium that Argentina obtained in the early 1980s. See L. S. Spector, with J. R. Smith, *Nuclear Ambitions* (Boulder, Colo.: Westview Press, 1990), 393–4 n. 58.

In early May 1991, apparently in response to behind-the-scenes US diplomatic interventions, China and Algeria announced that the Ain Oussera facility would be placed under IAEA supervision.[32]

The two countries also revealed that the reactor had a maximum power rating of only 15 megawatts and would normally be operated at 10 megawatts. Although large for a developing country, such a facility is not demonstrably inappropriate for peaceful nuclear research and the production of medical isotopes, and, even if operated at peak power, the unit would take one and a half to two years to produce enough plutonium for a clandestine nuclear device—assuming IAEA safeguards could be circumvented without arousing suspicion. These factors significantly reduced the risk that Algeria might build up a substantial stockpile of plutonium that could be quickly transformed into cores for nuclear weapons.

China and Algeria also disclosed that the agreement for the purchase of the reactor had been signed in February 1983. This preceded China's 1984 pledge to require safeguards on all nuclear exports. Chinese spokesmen implied that this timing justified the absence of any earlier notification to the IAEA about the project.

During the early 1980s, Algeria had announced discussions on a range of nuclear construction projects with several foreign suppliers, including France, Belgium, Brazil, and the United States. Algeria had also revealed tentative plans to build a research reactor at Ain Oussera. No agreements to build nuclear-research or power-reactor projects were known to have been concluded, however, until the 1985 reactor-purchase agreement with Argentina, and discussions with China remained undisclosed.[33]

Despite the decision to place the Ain Oussera reactor under safeguards and the reassuring declarations as to its size, a number of questions about Algeria's intentions in acquiring the facility remain unresolved. Most importantly, if the reactor was intended for peaceful purposes, it remains unclear why it was built in secret, why it was so heavily defended, and why China concealed

---

[32] 'Remarks by Spokesman of Chinese Foreign Ministry', 30 Apr. 1991 (courtesy, Embassy of the People's Republic of China); R. J. Smith, 'Algeria to Allow Eventual Inspection of Reactor, Envoy Says', *Washington Post*, 2 May 1991.

[33] Graham *et al.*, 'Nuclear Power and Nuclear Proliferation in the Islamic Middle East'.

its existence throughout the 1980s. During this period, Beijing held many rounds of talks with Washington in which China voiced its commitment to strict nuclear-export policies, but apparently never noted the Algerian reactor sale. It is also unclear whether China and Algeria planned from the start to place the facility under IAEA safeguards, or whether they took this step only to deflect criticism in the wake of the unit's discovery. Similarly, while the IAEA will be able to verify that the reactor is operated at 15 megawatts or less, doubts remain as to whether Algeria had originally planned to run it in this way. The facility's oversized cooling system suggests that Algeria had intended to operate the unit at a higher power level or to enlarge it in the future.[34] Indeed, an early 1981 press report stated that Algeria was seeking to acquire a reactor similar to the larger Osiraq unit.[35]

Algeria's possible motives for acquiring a nuclear facility with a military potential also remain unclear.[36] Algeria had been engaged in a seven-year guerrilla war with Morocco over Western Sahara when, in 1982, Rabat concluded a military-base agreement with the United States. Algeria may have feared that US military power would back Moroccan regional political ambitions and that an Algerian nuclear-weapons option could counter such interference. In the early 1980s, when it signed the contract with China, Algiers may also have seen the facility as bolstering its prestige, as it pressed its bids for leadership in the Maghreb and in the Non-Aligned Movement. The nuclear ambitions manifested by Libya and Iraq and the establishment of their respective nuclear-research centre at Tajoura and Tuwaitha may also have stimulated Algerian interest in acquiring the rudiments of a nuclear capability.

[34] Underlying these concerns as to the reactor's intended size were two historical antecedents. In 1960 Israel had claimed that its Dimona reactor was only a 26-megawatt plant, but in fact it appears to have been considerably larger. There was a similar controversy over whether Iraq's IAEA-safeguarded Osiraq reactor was a 40- or a 70-megawatt unit. An additional question, finally, is why Algeria needed the Chinese facility for research and medical-isotope production, when it already had the 1-megawatt research reactor supplied by Argentina that could be used for these purposes.

[35] 'Algeria Looks to Go Nuclear', 8 Days, 28 Feb. 1981, p. 46, cited in Graham et al., 'Nuclear Power and Nuclear Proliferation in the Islamic Middle East'.

[36] The author is indebted to Dr Robert Mortimer, Professor of Political Science, Haverford College, Haverford, Pa., for sharing his views on Algeria's strategic outlook in the early 1980s.

Changes in Algeria's political and strategic situation—including an abortive programme of democratization and economic reform, as well as the unravelling of a UN-sponsored settlement of the Western Sahara War—could increase such pressures as may exist in Algeria for the acquisition of nuclear arms as symbols of national strength.

In June 1991 unrest by Muslim fundamentalists led Algerian President Chedli Benjedid to postpone the country's first multiparty parliamentary elections since independence in 1962 and to widen participation in the ruling Revolutionary Command Council. The elections, which had been set for late June, were rescheduled.

In Algeria's December 1991 parliamentary elections, however, Muslim fundamentalist parties won by a landslide, raising concerns that the country's nuclear programme might soon rest with leaders more militant than President Chedli Benjedid and his reformist allies. This prospect contributed to intensified pressure on Algiers to provide assurances that its nuclear programme would remain entirely peaceful, and, on 7 January 1992, the Benjedid government unexpectedly announced that Algeria would join the NPT.

On 11 January, however, five days before run-off elections that were expected to enlarge the fundamentalists' parliamentary majority and establish conditions for Algeria's transformation into an Iranian-style Islamic republic, Benjedid resigned. By prearrangement, an anti-Islamic caretaker government, backed by the Algerian military, immediately took power. The action left the country's political future—and the future of its nuclear programme and NPT pledge—in considerable doubt.

ISRAEL

Public estimates of Israel's nuclear capabilities have been based in large part on analyses of the revelations made by Mordechai Vanunu to the London *Sunday Times* in October 1986.[37] A book

---

[37] 'Revealed: The Secrets of Israel's Nuclear Arsenal', *Sunday Times* (London), 5 Oct. 1986. Vanunu, a nuclear technician who had worked at Israel's classified Dimona nuclear-research centre between 1977 and 1985, was abducted by Israeli intelligence agents just before the *Sunday Times* story was published. See T. L. Friedman, 'Israeli Suspect Flashes a Hint He was Abducted', *New York Times*, 23

by Seymour Hersh, a highly regarded American investigative journalist, states, however, that Israel's arsenal is considerably larger and more advanced than these assessments have suggested, however.[38]

Based on the data supplied by Mordechai Vanunu, the *Sunday Times* projected that Israel might have as many as two hundred nuclear devices. US officials who attempted to harmonize Vanunu's testimony with other relevant information, however, concluded that Israel's nuclear armoury probably contained fewer than a hundred weapons, and perhaps no more than fifty or sixty. They believed that the Dimona reactor, though it may have been enlarged somewhat in the late 1970s, was simply not powerful enough to have produced the plutonium needed for such a large arsenal.[39] Either stockpile would permit Israel to use a number of its nuclear weapons tactically (i.e. against military targets) during a conflict, while keeping a number of weapons as a strategic reserve to threaten enemy cities.

Vanunu also provided detailed information indicating that Israel had produced tritium and lithium deuteride, suggesting that the Israeli arsenal consists, at least in part, of advanced nuclear weapons. It has been assumed that the tritium is used in the manufacture of 'boosted' nuclear weapons, nuclear devices that rely principally on the fissioning of plutonium or highly enriched uranium for their yield, but which use the fusion of a small amount of tritium to enhance the fission process.[40] In practice, this means that some Israeli nuclear weapons may have yields

Dec. 1986; 'How Israeli Agents Snatched Vanunu', *Sunday Times* (London), 9 Aug. 1987; 'Riddle of Vanunu Ship', *Sunday Times* (London), 16 Aug. 1987; 'Revealed: The Woman from Mossad', *Sunday Times* (London), 21 Feb. 1988. (The Israeli government has never explained how it obtained custody of Vanunu.) Vanunu was subsequently tried and convicted of espionage and treason, receiving an eighteen-year prison term (see G. Frankel, 'Israel Convicts Vanunu of Treason for Divulging A-Secrets to Paper', *Washington Post*, 25 Mar. 1988). Israel's actions indicate that his revelations were considered a major security leak and tend to confirm the authenticity of his disclosures.

[38] S. M. Hersh, *The Samson Option* (New York: Random House, 1991).

[39] Although Vanunu observed large quantities of plutonium being separated while he worked at the Dimona plutonium separation plant between 1977 and 1985, the US specialists assume that the plant was working off an accumulated backlog of nuclear material from the Dimona reactor and that afterwards, the output of the plutonium plant was reduced to match the output of the reactor.

[40] After detonation, the tritium undergoes an atomic fusion reaction, releasing a stream of neutrons to boost the efficiency of the fission reaction, allowing a greater nuclear yield.

several times greater than the 20 kilotons nominally assumed to be the size of developing-country nuclear devices.

In addition, former US nuclear-weapons designer Theodore Taylor, after examining Vanunu's photo of an unusually configured nuclear-weapon component made from lithium deuteride, suggested that Israel may also be building 'super-boosted' weapons with a yield of perhaps 100 kilotons, the size of some warheads in US strategic missiles.[41] Taylor and others have assumed that the manufacture of full-fledged, multi-stage thermonuclear 'hydrogen' bombs with megaton yields[42]—or equally complex low-blast, high-radiation 'neutron' bombs—would require extensive full-scale nuclear testing and thus remain beyond Israeli abilities, since Israel is not known to have conducted the necessary tests. It has also been widely assumed that nuclear weapons with small physical dimensions, such as artillery shells, land mines, and 'suitcase' bombs, would similarly have required a nuclear testing programme.[43]

There has been no conclusive proof that Israel has ever conducted a full-scale nuclear test. Its nuclear arsenal is thought to have been developed, in part, through the testing of non-nuclear components and computer simulations—and through the acquisition of weapons design and test information from abroad.

Israel is thought to have obtained data from France's first nuclear test in 1960.[44] It may also have obtained data from US nuclear tests at approximately the same time. According to a May 1989 US television documentary, Israel was able to gain access to information concerning US tests from the 1950s and early 1960s. The test data could have included the results of US boosted and thermonuclear weapons that were being developed at the time.[45]

[41] A kiloton is the equivalent of 1,000 tons of TNT.
[42] A megaton is the equivalent of 1 million tons of TNT.
[43] Vanunu told British physicist Frank Barnaby that, in 1977, Israel began producing small quantities of lithium-6, a key precursor for the production of both tritium and lithium deuteride. Full-scale production of lithium-6 began in 1984 and continued for two years; tritium production (and presumably lithium-deuteride production) also began the same year. Altogether, Vanunu said, Israel produced approximately 375 lb of lithium-6, sufficient for a sizeable arsenal of advanced nuclear devices.
[44] 'France Admits it Gave Israel the A-Bomb', *Sunday Times* (London), 12 Oct. 1986; S. Weissman and H. Krosney, *The Islamic Bomb* (New York: Times Books, 1981), 114.
[45] 'Israel: The Covert Connection', Frontline, Public Broadcasting System, 16 May 1989.

In addition, there has been speculation that a signal detected on 22 September 1979, by a US Vela monitoring satellite, over the south Atlantic, was, in fact, the flash from a low-yield Israeli nuclear test—possibly of a tactical nuclear weapon or of the fission trigger of a thermonuclear device.

Seymour Hersh's book concludes that Israel's nuclear capabilities are considerably more advanced than the foregoing analysis suggests. Relying largely on interviews with US intelligence analysts and Israeli figures knowledgeable about the country's nuclear programme, Hersh claims that Israel now possesses 'hundreds' of low-yield, neutron-bomb-type warheads, many in the form of artillery shells and land mines, as well as full-fledged thermonuclear weapons.[46] He also states that Israel deployed some of the land mines in the Golan Heights in the early 1980s.[47]

To help explain how Israel's nuclear weapons could be so advanced, Hersh states that Israel had access to the information from numerous French nuclear tests. He also reports that, 'according to Israeli officials whose information about other aspects of Dimona's activities has been corroborated', the September 1979 event observed by the US Vela satellite in the south Atlantic was indeed an Israeli test—and was actually the third of a series of tests conducted at that time.[48] The first two tests, Hersh's sources state, were obscured by storm clouds. The claim that clouds would prevent detection of an atmospheric nuclear detonation by a Vela satellite has been challenged, however, since the satellite is said to rely in part on infra-red sensors that can penetrate cloud cover. Thus this critical matter remains unresolved.

---

[46] Hersh, *Samson Option*, 291, 312, 319. Hersh states that analysts at the Lawrence Livermore National Laboratories determined that Israel had produced neutron bombs from their examination of Vanunu's photographs. He also states that, in 1981, another scientist or engineer who had worked at the Dimona complex provided the United States with photos of a nuclear-weapons storage bunker that indicated Israel possessed hydrogen bombs. The defector told US intelligence analysts that Israel had more than a hundred weapons in storage at the time. This, however, would have been several years before Vanunu said that Israel began full-scale production of lithium-6—needed for the production of lithium deuteride for hydrogen bombs—raising a series of new questions.

[47] Such an action would be quite dangerous, in that it might lead to nuclear escalation early in a Syrian–Israeli conflict, since, in the event of a sudden Syrian thrust into Israeli-held territory, Israel would be compelled to use the weapons or lose them to its foe.

[48] Hersh, *Samson Option*, 271.

By 1973, Hersh reports, Israel possessed twenty nuclear devices, including some small enough to fit into a suitcase. At this point, he states, Israel also possessed missiles capable of hitting the southern Soviet Union—suggesting that the early version of Israel's Jericho missiles had a range close to eight hundred miles, not the four hundred miles previously believed. By 1977, Hersh recounts, Israel possessed well over a hundred nuclear weapons and expected to have two hundred by 1980, according to intelligence obtained by the United States at the time from a knowledgeable Israeli. The timeframe for the rapid increase in Israeli capabilities overlapped Vanunu's tenure at the Dimona plutonium-separation plant, which, according to Vanunu, was then operating at more than 100 per cent of its design capacity. None the less, Hersh's figures suggest a far more rapid growth of the Israeli nuclear arsenal than that found in previous analyses.

Unfortunately, Hersh does not provide any technical data to support his claims as to the size and sophistication of Israel's nuclear armoury. Thus, basic questions remain as to how Israel could have obtained the necessary weapons-grade material to build the capability he describes.[49]

For this reason, additional substantiation will be required before Hersh's expansive vision of Israel's nuclear capabilities can be fully accepted. None the less, Hersh's respected reputation for unearthing secrets and the multiple sources he cites lend credibility to his claims. At a minimum, military strategists in other Middle Eastern states will have to incorporate Hersh's assessments about Israel's nuclear might into their 'worst-case' planning assumptions.

---

[49] Israel is thought to have the ability to enrich uranium, and perhaps this has been used to augment its production of material for nuclear arms. Testimony by Director of Central Intelligence William Webster in 1989, however, implied that Israel did not have a significant enrichment capacity, and Vanunu—who, in any event, had only second-hand knowledge of Israel's enrichment programme—stated that the country began enriching uranium only in 1979, which would mean that enriched uranium could not have contributed to the rapid increase in the Israeli nuclear arsenal before that time. See F. Barnaby, *The Invisible Bomb* (London: I. B. Taurus, 1989), 25; Testimony of William Webster, Director of Central Intelligence, before the Senate Governmental Affairs Committee, *Hearings on Missile and Nuclear Proliferation*, US Senate, 101st Cong. 1st Sess., 18 May 1989. Hersh notes Israel's enrichment capacity only in passing, apparently relying on Vanunu, and does not argue specifically that it has contributed substantially to the country's nuclear might.

## THE NON-PROLIFERATION TREATY

Iraq's deliberate efforts to circumvent the NPT, and the inability
of the IAEA to detect these violations, have shaken confidence in
the accord as a mechanism for constraining the nuclear activities
of signatory states. Particularly disturbing is that some of these
violations would have remained undetected but for the 1991
Gulf War and might eventually have permitted Iraq to acquire a
number of nuclear weapons without being observed.

As noted earlier, parties to the treaty are required to place all of
their nuclear materials and the facilities that process them under
IAEA inspection to ensure that they are not diverted for military
purposes. In the past, the IAEA has relied on the inspected state to
declare all of its nuclear activities and has then limited its inspections to declared material and facilities.

Iraq's most serious and repeated NPT violation was the construction and operation of undeclared facilities. Even if the
existence of Iraq's undeclared facilities had been known, however,
the IAEA had few tools to deal with them, since it has traditionally
confined its safeguarding operations to declared installations.
While IAEA agreements with NPT parties authorize the agency to
undertake 'special inspections' of suspected undeclared nuclear
sites with the consent of the inspected country, the agency had
never attempted to exercise this authority.

In the wake of its setbacks in Iraq, however, the organization
began to shed its timidity. In late 1991 the agency's Board of
Governors revitalized the agency's special inspection rights. In
addition, IAEA Director General Hans Blix has established a unit
to receive and evaluate intelligence, to be supplied by the United
States and other IAEA members, about suspected covert facilities
and other NPT violations. The IAEA Board also set up expedited
procedures for involving the UN Security Council, in the event
that a target country refuses a special inspection or is found to
have violated IAEA rules.

These initiatives should greatly strengthen the IAEA's hand in
deterring violations similar to Iraq's by other NPT parties in the
Middle East. A first test of the fortified inspection system is taking
place in Iran. The location of its weapons-related nuclear research-
and-development activities remains uncertain, but several sites are
under suspicion, including one north-west of Tehran and another

near Qazvin.[50] For obvious reasons, Iran has not declared any of these sites to the IAEA. If facilities at any of these locations are producing or using nuclear materials, or if they are being built to do so, Iran is violating its IAEA obligations by not allowing agency inspectors to scrutinize the installations. In February 1992 the IAEA visited several such suspect sites at the invitation of the Rafsanjani government. Following the visit, the IAEA announced that it had found no evidence of a clandestine nuclear-weapons programme at the locations it observed. The agency's credibility was marred, however, by accusations that it had been misled into visiting an alternative location with the same name as one of Iran's allegedly undeclared nuclear facilities. With continuing suspicions in the international community as to Iran's nuclear intentions, it remains to be seen whether the agency will request further special visits to validate its initial findings.

The agency's expanded safeguards authority, if credibly used, could also be helpful in deterring Libya from pursuing a covert nuclear-weapons programme. Although its nuclear programme appears to have been dormant during the latter half of the 1980s, Libya succeeded in building one of the world's largest chemical-weapons plants through clandestine purchases of equipment and material in Western Europe. This experience might tempt it to use the same strategy to advance its nuclear ambitions.

The strengthened IAEA system could serve as an important constraint on Syria, as well, at a time when Damascus is beginning to pursue the development of a nuclear infrastructure with suspect objectives.

In Israel and Algeria, neither of which is a party to the NPT, the IAEA's activities are confined to individual facilities that these states have voluntarily agreed to place under agency inspections. Special inspections of suspected undeclared nuclear sites are not contemplated, since these countries, by remaining outside the NPT, retain the right to build and operate uninspected facilities. As in the case of comprehensive NPT-based safeguards, however, the IAEA has recourse to the Security Council to enforce the more limited inspection rights that do apply in these instances.

In Israel, only one facility is currently under IAEA inspection,

[50] Aldinger, 'Iran not Close to Developing Nuclear Arms'; 'The China–Iran Nuclear Cloud', *Mednews*.

the small Nahel Soreq research reactor, which has never been linked to Israel's nuclear-weapons programme. Algeria's intended use of the Ain Oussera reactor is more problematic, however. If the IAEA establishes the precedent of vigorous implementation of its special inspection rights in NPT states and strengthens its ties to the Security Council to enforce these rights, these steps will undoubtedly enhance the value of its safeguards as a deterrent against cheating at inspected installations—such as the Ain Oussera facility—in non-NPT countries. Thus, except in the notable case of Israel, the impending improvement of IAEA monitoring could have an important impact throughout the region.

To be effective, special inspections ultimately depend on accurate intelligence concerning undeclared nuclear activities. In Iraq, such intelligence was lacking, and, even if the IAEA had been prepared to demand a special inspection, it would not have known where to look, in many cases. The failure of US intelligence to detect the construction of the Ain Oussera reactor for a number of years is also cause for concern. With the end of the Cold War and the disintegration of the Soviet Union, the United States has greatly increased its commitment of intelligence resources to the issue of proliferation, which may go far towards addressing this problem.

The particulars of the special-inspection regime also deserve close scrutiny. If the threshold of proof for initiating a special inspection is too high, the technique may prove unusable as a practical matter. It is also important that the time between the demand for a special inspection and the arrival of inspectors at the target site be sufficiently short to prevent practices of the type Iraq repeatedly employed to deceive the UN–IAEA inspection teams operating under Resolution 687.

In sum, IAEA safeguards are being significantly improved and could have an important restraining impact on a number of Middle Eastern nuclear programmes of proliferation concern.

## CONCLUSION

Although it is highly probable that Israel will retain its nuclear monopoly for much, if not all, of the 1990s, the nuclear map of the Middle East is changing dramatically. Developments in Iran, Syria, and Algeria leave no doubt that the nuclear aspirations of other regional states are intensifying, and concerns remain about a future resurgence of Iraq's nuclear-weapons effort. Ironically, as international attention is focused on the Bush Administration's Middle East Arms Control Initiative and Egyptian President Hosni Mubarak's plan for a regional zone free from weapons of mass destruction, the most potent instrument for restraining the spread of nuclear weapons to new states may well prove to be the very one—in strengthened form—that failed so utterly to halt Iraq's clandestine bid for the bomb.

# PART THREE

# STRATEGIC AND POLITICAL IMPLICATIONS

# 7

# The 'Proliferation Problem' and the New World Order

## LAWRENCE FREEDMAN

Two events pushed the proliferation question to the fore during the early 1990s. One was alarm that the fragmentation of the Soviet state would lead to the fragmentation of its nuclear arsenal. The other was the set of revelations at the time of the 1991 Gulf War and its aftermath concerning Iraq's actual acquisitions of both chemical weapons and ballistic missiles and, most seriously, its prospective acquisition of nuclear weapons. Western states were given a jolt. In contrast with earlier proliferation scares—such as, for example, those involving the Indian subcontinent—Western states could see direct security implications for themselves in both these developments. They did not simply complicate regional politics or set a bad example to others, but threatened the management of the international system—just at the point when other developments appeared to have made this management much easier.

### THE PROLIFERATION PROBLEM

Proliferation is often discussed simply in terms of being undesirable in its own right rather than in relation to its effects on the international system as a whole or, more narrowly, on Western interests. The problem is seen as stemming from a dangerous combination of technical opportunity, as a function of the diffusion of the relevant technology, with mischievous motivations—hence those numerous edited collections which combine essays on the nuclear-fuel cycle and proposals for tightening the non-proliferation regime with analyses of the state of play of the policies of the usual suspects.[1]

For those opposed to all nuclear arsenals, one state's nuclear capability is as bad as another's. Mainstream strategists, however, have taken a more nuanced line. Nuclear weapons have been seen as performing a valuable strategic service for the West, but as a potential hindrance to the development of sensible security policies elsewhere. This appears to be because doubts over the political stability of potential proliferators make it seem more likely that their weapons might actually be *used*. However, there is no proof that only additional powers pose a danger in this regard. The Soviet Union, after all, was once considered to be a paragon of stability if nothing else and thus a responsible custodian of a nuclear arsenal. Even if the number of nuclear powers had been held at three or five, the Soviet nuclear legacy would still be causing headaches today. Equally, as Kenneth Waltz has argued, there is no reason in principle why other groups of states cannot set up systems of mutual deterrence comparable to the one which is credited with keeping the peace in Europe during the Cold War period.[2]

It is possible to design international structures which are reinforced by nuclear arsenals and others which may be torn apart. In the former case, weapons may become familiar pieces of furniture, helping to sustain common sense by acting as symbolic reminders of the folly of total war: in the latter case, the 'last resort' may arrive and weapons kept in reserve for such a moment will be unleashed. Getting the measure of the proliferation problem, therefore, requires an appreciation of its shifting international context. Tracing the relationship between the definition of the problem and the international system makes it possible to explore the elements of continuity and change in both.

[1] For some examples, see R. M. Lawrence and Joel Larus (ed.), *Nuclear Proliferation: Phase II* (Lawrence, Kan.: University Press of Kansas, 1974); G. Quester (ed.), *Nuclear Proliferation: Breaking the Chain* (Madison, Wisc.: University of Wisconsin Press, 1981); J. Snyder and S. F. Wells, jr. (eds.), *Limiting Nuclear Proliferation* (Cambridge, Mass.: Ballinger, 1985); J. Goldblat (ed.), *Non-Proliferation: The Why and the Wherefore* (London: Taylor & Francis for SIPRI, 1985); P. Lomas and H. Muller (eds.), *Western Europe and the Future of the Nuclear Non-Proliferation Treaty* (Brussels: Centre for European Policy Studies, 1989).
[2] K. Waltz, *Spread of Nuclear Weapons: More may be Better* (Adelphi Papers, 171; London: IISS, 1981).

In this chapter I will argue, through a brief review of this issue during the post-war period, that the critical variable is the prevailing alliance structure. Broadly speaking, even a quite loose alliance with an established nuclear state is an alternative to proliferation. Drawing on the deterrent power of another may carry fewer risks as well as lower costs than a drive for a national capability. The incentives for proliferation grow with the lack of a reliable superpower protector. I will also argue that the current stage in international affairs—the new world order—may well involve a weakening of alignments. If so, then that suggests a potential aggravation of the proliferation problem.

## ALLIANCES AND PROLIFERATION

Speculation on the spread of nuclear weapons began immediately after Hiroshima. Initially, 'outsiders' tended to think ahead to a world of nuclear powers. Some saw this in apocalyptical terms, imagining wars opening with devastating blows from an unknown enemy, or, more optimistically, a developing system of mutual deterrence, thus anticipating later debates.[3] In the United States the 'insiders' at least preferred to think of an extended US monopoly and so followed a resolute non-proliferation policy. This included not only the Soviet Union but also the United Kingdom, despite the fact that the bomb itself was a product of wartime co-operation. The 1947 McMahon Act shut the United Kingdom out of the US programme. Before that, the Baruch Plan at the United Nations envisaged international management of civil nuclear energy. From Moscow's perspective, the purpose of the plan was judged to be the maintenance of a US monopoly. Under its terms, the Soviet programme would have been aborted at once, while the US programme was left in place until the very last stages of the plan. It was not hard to imagine Washington finding a pretext for postponing the final elimination of its own arsenal.

In non-proliferation terms, this approach was a complete failure. Both the Soviet Union and the United Kingdom pressed ahead with their own programmes, which soon reached fruition.

---

[3] See e.g., J. Viner, 'The Implications of the Atomic Bomb for International Relations', *Proceedings of the American Philosophical Society*, 90/1 (Jan. 1946), repr. in P. Bobbit, L. Freedman, and G. Treverton, *US Nuclear Strategy: A Reader* (London: Macmillan, 1989), 54–63.

The strategic consequences of the Soviet capability were profound: over time it neutralized the key US advantage and made it much easier to consolidate Moscow's hold on Eastern Europe. The UK programme was less important. The Eisenhower Administration acknowledged that it was odd to refuse to co-operate with the United Kingdom on all aspects of NATO other than nuclear weapons at a time when its strategy was based on deterrence. It therefore agreed to pool Anglo-American efforts. This eventually turned into a significant subsidy for the UK programme.

The next major anti-proliferation effort was the 1954 'atoms-for-peace' initiative, which sought to head off pressures to develop military nuclear programmes by offering assistance with civilian nuclear programmes. It was not fully realized until later that, by providing a mechanism to distribute basic technology and facilities, this was probably the single most important stimulus to proliferation after the Smyth Report—the 'how we did it' document produced after Hiroshima which helped the Russians as much as the 'atomic spies'.[4]

In strategic terms the debate on proliferation became bound up with questions of alliance management and was described as the 'Nth country problem'.[5] By the end of the 1950s nuclear weapons were seen as the means by which first-class states were distinguished from the rabble and also the best guarantee of national security. If a number of Western states acquired a nuclear capability, then that would compromise the US position as head of the alliance: it would only be one of a number of 'first-class' states, while attempts to impose discipline in dealings with the Warsaw Pact would be undermined if recalcitrants knew that they could rely on their own deterrents should a break with Washington become necessary. Indeed, in the most impressive pro-proliferation tract of

[4] H. D. Smyth, *A General Account of the Development of Methods of Using Atomic Energy for Military Purposes under the Auspices of the US Government, 1940–45* (Washington DC: US Government Printing Office, Aug. 1945). For an early critique of the consequences of Atoms for Peace, see A. Kramish, *The Peaceful Atom in Foreign Policy* (New York: Council on Foreign Relations, 1963).

[5] A. Wohlstetter, 'Nuclear Sharing: Nato and the N + 1 Country', *Foreign Affairs*, 34/3 (Apr. 1961). F. Ikle ('Nth Countries and Disarmament', *Bulletin of the Atomic Scientists*, 16/10 (Dec. 1960), 391–4), took a more sceptical view. See also R. N. Rosecrance (ed.), *The Dispersion of Nuclear Weapons* (London: Columbia University Press, 1974). The first book-length discussion of the problem was L. Beaton and J. Maddox, *The Spread of Nuclear Weapons* (London: Chatto & Windus, 1962).

the time, by Pierre Gallois, the basic contention was that alliances were untenable in the nuclear age. It was too much to expect one state to put itself at nuclear risk for another and so the only sure basis for security was a national nuclear force.[6]

At stark moments of truth this might be the case, but when the threats to security were less than ultimate then alliance still held many advantages. General de Gaulle's advocacy of a French nuclear programme had less to do with a strategy for survival than the political consequences of dependence upon the United States. His alternative vision for Europe was one in which US hegemony had been weakened sufficiently for France, having regained its stature of a power of the first rank, to play a leadership role of its own.[7] No other Western state accepted the view that its security would benefit significantly from the erosion of the Atlantic Alliance, and so France became isolated within NATO. Because of this, it was forced, especially after de Gaulle's departure, to repair its relations with its allies. At the same time NATO doctrine was adjusted to take account of the potential deterrent value of the UK and French forces, most notably in the 1974 Ottawa declaration.[8] The concept of multiple decision centres which helped rationalize this shift played on potential incoherence in the alliance at times of extreme stress—Moscow would have to form a judgement not only on whether Washington would accept a risk of national suicide on behalf of Europe, but also on the calculations of London and Paris. The strategic logic can be taken with a pinch of salt. The value of the doctrine was to help accommodate and contain proliferation within an alliance framework.

The Soviet Union was less successful in its relations with its most awkward ally. China, having been the recipient of direct nuclear threats from the United States during the Korean War, had more reason to feel insecure. A combination of weak backing during the offshore-islands crisis of 1958 and then the withdrawal of all nuclear assistance convinced Mao Zedong that Moscow was an unreliable partner. There are many reasons why a Sino-Soviet split was likely in the early 1960s, but arguments over the limits of nuclear protection certainly played a key role as a catalyst.

[6] P. Gallois, *The Balance of Terror: Strategy for the Nuclear Age*, trans. by Richard Howard (Boston: Houghton Mifflin, 1961).

[7] W. Mendl, *Deterrence and Persuasion* (London: Faber & Faber, 1970).

[8] Declaration of North Atlantic Council, Ottawa, 19 June 1974.

The moves towards both a test ban and a non-proliferation treaty were seen—with some justice—by France and China as being directed against their nuclear programmes and for this reason they chose not to co-operate, though in the end the NPT consolidated their nuclear status. By the time the treaty was ratified in 1970 the strategic debate on non-proliferation had shifted. At issue now was the position of states outside the established alliance structures.

The number of such states had grown through the 1960s. Although the more problematic states in proliferation terms tended to be among those who had been among the first to achieve independence after the war, such as India and Israel, the context was one in which the sheer number of newly independent states was coming to compromise bipolarity.

At the start of the 1950s the Cold War dominated both European and east Asian politics. However, the attempt to extend the alliance system through south Asia and the Middle East during the decade indicated the problems of attempting to squeeze the politics of these areas into the simple 'two-camp' philosophy adopted by both superpowers. By the 1960s bipolarity was being questioned and qualified as the two superpowers modified their antagonism. Their evident determination to avoid a head-on collision and the priority being given to the 'management' rather than the resolution of crises were among the reasons why France and China felt that they no longer had a reliable security guarantee.

The non-aligned movement reflected the conviction of the newly independent states that they had little to gain by being forced into either the Western or the Eastern camp. The logical corollary of non-alignment was the creation of independent military capabilities. For those still locked in a 'national-liberation' philosophy, this was best done through forms of guerrilla warfare, which were seen to put the Western powers at a severe disadvantage. For those able to look beyond the liberation struggle, the natural course was to build up military capabilities to match those of the established powers. China followed both paths. Again this was often as much a question of political influence as military doctrine. India, which was the most articulate exponent of this view, saw nuclear status as helpful in its conflicts with both China and

Pakistan but also as a means of reinforcing its case for membership of the international élite.[9]

By 1974, when India tested its first nuclear 'device', there was declining confidence in the relevance of bipolarity. A distinctive Third World philosophy was taking shape, associated with the Group of 77, whose chosen battleground was the United Nations Conference on Trade and Development. The confidence of its members (though not generally their economies) was boosted by the success of OPEC. At a time when East–West *détente* seemed to be taking root, it soon became almost a cliché that the dominant lines of antagonism were now North–South. The old alliances seemed to be on the verge of breakdown. China and the Soviet Union were at loggerheads, and on occasion close to war, while members of the Warsaw Pact could only be held down through 'fraternal' interventions. On the Western side the deterioration was not so marked, but there was none the less an awareness that the US monopoly of power was being challenged in the economic sphere if not the military by Europe and Japan.

Although the 1980s opened with a revival of Western insecurity as a result of the aftermath of the overthrow of the Shah in Iran and the Soviet intervention in Afghanistan, by the end of the decade there were grounds for satisfaction that none of the worst fears had been realized. The Soviet Union was being consumed by its economic failings. It had become irrelevant to most Third World states, except when providing arms and military advisers, neither of which appeared to be sufficient to prevail in local conflicts, while none could ignore Western economic strength. OPEC's greed in the 1970s had forced the West into energy savings and alternative sources of supply and so there was a glut. Two of the key radical states in the Middle East—Iraq and Iran—were at war with each other. Instead of the Left supporting insurgencies, the United States was backing guerrilla campaigns by the Contras in Nicaragua, UNITA in Angola, the Mujahïdin in Afghanistan.

These tendencies ran their course in the 1990s with the complete collapse of communism and the comprehensive defeat of a

[9] For an Indian critique of the NPT, see K. Subrhmanyam (ed.), *Nuclear Proliferation and International Security* (Delhi: Institute for Defence Studies, 1985).

Third World regime when it dared to take on a Western-led coalition. One of the most striking shifts took place in the United Nations, which moved from empty resolutions in the General Assembly dominated by Third World rhetoric to Western-led initiatives in the Security Council.

## PROLIFERATION AND THE INTERNATIONAL POWER STRUCTURE

What is striking about the many changes that took place in international affairs after 1945 is how irrelevant nuclear weapons have been except in the admittedly critical area of East–West relations. Despite all the 1970s talk of multipolarity, they did not appear as a means of marking out a new international power structure. The idea of a 'Euro-bomb' was discussed regularly but always in the future tense, as the logical conclusion of a process that was not yet properly under way. France saw its nuclear capability as raising its own national standing rather than that of Europe as a whole. The United Kingdom's nuclear links were with the United States and it was ultra-loyal to NATO. Germany, having toyed with the idea of a nuclear programme in the late 1950s, was well aware of the anxieties this would arouse, while Japan, as the only victim of nuclear attack, maintained its 'nuclear allergy'. India's nuclear explosion did not propel it into a political super-league. Even China's much-vaunted attainment of super-power status, which seemed to many to be a natural consequence of its combination of vast population and nuclear weapons, did not materialize. Its economic weakness limited its ability to exert more than a regional influence.

Nuclear proliferation did not develop at anything like the pace anticipated in either the early 1960s or the early 1970s. The point was that, unless there was a compelling strategic case for proliferation, alternative arguments for a nuclear capability lost force. At times nuclear power as a symbol of modernization or as an alternative energy source to oil had attractions, but such arguments provided no case for a military option, and had anyway run their course by the early 1980s. Prestige arguments were undermined by the setting of an international non-proliferation norm by the NPT and by the fact that Japan and Germany, the two more successful powers of the 1970s and 1980s, were non-nuclear.

Prestige has served as an unconvincing rationale on its own, especially for countries with weak economies. It is notable that one of the few successes in recent years has been the reduction in the proliferation impulse in recent years as strategic arguments have lost credibility. The United Kingdom only started to employ 'seat-at-the-top-table' sorts of rationales for a national nuclear force when its 'added strategic value' became questionable. It was in close alliance with the United States, which could strike on its own any target that might need striking without UK assistance. To draw attention to the possibility that one day it might be necessary to 'stand alone' without US protection risked undermining NATO and the US nuclear guarantee to Europe. Even if the United Kingdom could look after itself, it was unlikely to be able to substitute for the United States in extending protection to all the NATO countries, so it did not want to put itself in the position of being asked to do so by casting doubt on the durability and sincerity of US commitments. Hence rationales came to revolve around the influence that it could wield as a nuclear power, coupled with vague nods to more drastic scenarios in references to 'ultimate guarantees of national security'. This was good enough for a country which had been in on the start of the nuclear age and with an advanced nuclear capability subsidized by the United States: it would have been unlikely to justify the necessary investment if starting from scratch.[10]

The most compelling arguments for the acquisition of a nuclear arsenal are strategic. Not surprisingly, therefore, proliferation has come to be associated with a sort of international underclass—'pariah' states, which either felt under threat and uncertain of any outside support, or else were run by militarist regimes. South Africa and Israel were put in the former category; Argentina and Brazil in the other. In one sense, pariah status became a self-fulfilling prophecy after the NPT. Any state which broke the non-proliferation norm risked being 'blackballed', as India found in 1974. India was large enough to withstand international indignation. Smaller states were not. Moreover, an announcement of nuclear ambitions from one state meant that other states which had reason to feel threatened might set their own programmes in

[10] L. Freedman, M. Navias, and N. Wheeler, *Independence in Concert: The British Rationale for Possessing Strategic Nuclear Weapons* (Nuclear History Program, Occasional Paper 5; College Park, Md.: Center for International Studies, University of Maryland, 1989).

motion—or, if they were already nuclear, might consider pre-emption. Thus China's 1964 explosion led to a tense few years with the Soviet Union and also spurred on India's programme. Meanwhile India's own explosion resulted in Pakistan giving a higher priority to its own drive for a bomb.

One consequence of this was to slacken south Asian relations with the major powers. India and Pakistan could have appeared as Cold War proxies, but in practice India's relationship with the Soviet Union was always loose, while Pakistan's with the United States was never as close as implied by formal alliance and was later weakened by the attempt to stop Pakistan from going nuclear. Evidence that an Asian conflict had a higher chance than most of seeing nuclear weapons in use became a good argument for the major powers to stay clear themselves. In the Middle East the United States did turn something of a blind eye to Israel's bomb, but in return the Israelis took care not to boast. For Israel, US support was still preferable to a stand-alone policy. Again we see the link between patterns of proliferation and international align-ments. Proliferation is both symptomatic of—and responsible for—the loosening of strategic links. This helps explain perhaps why there is increasing concern about it in the 1990s.

## THE NEW WORLD ORDER

The fundamental changes to the international system consequent on the inner collapse of the Soviet Union and the discrediting of communism as a political philosophy are often summed up in the slogan 'new world order', which was used by President Bush to describe what he judged to be at stake during the Gulf conflict.[11] The starting-point of Bush's analysis was the end of the Cold War. This made possible co-operation with Moscow in the face of major challenges to international law. The United Nations could work as originally intended, and there would be no need to judge every regional issue in terms of its relevance for the superpower confrontation.

The language with which this opportunity was described was often over-exuberant and seemed to promise an impossibly com-

[11] L. Freedman, 'The Gulf War and the New World Order', *Survival*, 33/3 (May–June 1991), 195–209.

plete transformation in world politics. Inevitably, much of the old political game went on as before. The new world order could also be seen as an excuse for intervention in the internal affairs of other states and an attempt to refashion the world according to Western values. My interpretation of Bush's concept is that it was not a basis for getting too involved in Third World disputes, about which he was wary. The new world order was not so much a charter for universal human rights but a belief that the traditional rights of states could now be protected if the world's great powers acted together.

According to this approach, the key determinant of the old order was bipolarity. The key feature of the new system is not so much unipolarity, which suggests a hegemonic drive in the remaining superpower and in so doing overestimates both its willingness and ability to cast itself in these terms, but the end of a central, overriding strategic antagonism. As this is no longer an organizing principle for the major powers, they are better placed to keep an eye on the activities of the minor powers. Inevitably, great stress is placed on respect for the territorial status quo, on the grounds that, once historically based challenges to existing borders are allowed, then aggression becomes easier to justify and a single precedent could spark the revival of numerous ancient disputes.

There is little to object to in this approach, as far as it goes. The problem is that it pays insufficient regard to the other defining feature of the post-war international system—decolonization— which had begun to qualify bipolarity long before the European revolutions of 1989. In most regions of the world, decolonization has been more critical than the Cold War, for it has meant the disengagement of the old colonial powers, followed by the development of a complex and fluid local balance-of-power system. This is what has now happened to Europe. There is irony in this being the last continent to decolonize, given that it was responsible for the colonial period in the first place.

The decolonization process began as the Second World War ended, and so, in contrast to the Cold War, which became a source of stability in the system, it has provided a continual and largely destabilizing dynamic. The sense of abrupt change associated with the end of the Cold War is eased when the demise of the Soviet empire is seen simply as the natural culmination of the most profound trend in contemporary international history.

The rather classical ideas represented by the new world order only become relevant when local balances of power get seriously out of kilter and one state moves aggressively against another. They are less relevant for the generality of contemporary security problems, which are normally driven by very particular local rivalries, based on ethnic and religious and occasionally ideological divisions, interacting with state boundaries which were arbitrarily set in the colonial age and are often inappropriate for the social mix contained within them. There are very few true 'nation-states'. Most states contain a number of nationalities, while most nationalities straddle a number of states. In circumstances where there is no political authority able to transcend ethnic–religious divisions, then it is natural that authority will shift to those reflecting these divisions. The extent to which this might happen and the severity of the consequences depend in part, but only in part, on economic conditions.

These pressures, familiar in most regions of the world, are now starting to afflict Europe and are setting a new agenda. To what extent can economic assistance produce political stability? How seriously should the principle of self-determination be taken when its logic leads to demands for political units of limited viability? What are the optimum methods for picking up and rearranging the pieces of states which have irretrievably fallen apart? In what ways can international bodies help as conciliators in messy local disputes? At what point does external intervention become appropriate to protect beleaguered minorities or to keep the peace? Do states have rights to possess any weapons they choose, and, if not, why do some states have more rights than others?

The new-world-order approach suggests that standards of behaviour must be set by the politically stable and responsible states, with economic, diplomatic, and, where appropriate, military support conditional on these standards being met. The problem is that germane standards are often difficult to set. Contradictory principles are often at stake, while a pragmatic analysis of the consequences of imposing penalties for failing to meet approved standards warns that this may just aggravate the situation. In practice, therefore, the task of establishing codes of behaviour must go hand-in-hand with an attempt to construct a credible local balance of power, and this may in turn require significant external involvement. There is often the option of staying clear of hopeless conflicts. That can undermine the credibility of commitments to intervene elsewhere, but a more influential consideration may simply be that conflicts taking place in areas where Western interests are at stake cannot be easily ignored.

This creates a fundamental tension in Western policy-making. At one level there is a reluctance to engage in local conflicts until such times as they represent a clear threat to 'international peace and security'. This is backed by a renewed commitment to deal decisively with such threats when they do emerge. At another level there is a recognition that a degree of regular and intensive engagement is unavoidable if there is to be any real 'order' in the system and if Western interests are to be protected, but there is no agreed set of principles to back this up.

The tension is between the maintenance of general standards and the management of particular crises. This tension is not new. What is new is the extent to which this tension now dominates the immediate questions of European security, as well as the lack of a clear strategic imperative to provide an organizing principle for crisis management. The old Cold War imperatives did provide such a guide—albeit often of dubious reliability.

Attempts to identify a new strategic imperative have not been convincing. These often seem to represent attempts to turn local problems into global threats, often by exaggerating linkages between similar but largely unconnected practices, such as terrorism or drug-trafficking. Another approach is to see Islam as a new hostile force, largely by refusing to differentiate among its numerous varieties and political forms. Lastly, problems which have a global character, such as those of the environment, are turned into security issues even though they bear no relation to traditional concepts of security.

## REDEFINING THE PROBLEM

The simple new-world-order perspective helps to define the proliferation problem in terms of states developing aggressive options for themselves through the creation of irresistible local power and a capacity for deterring Western intervention on the side of the aggrieved. Hence the frequent discussion of the problem in terms of 'what if [insert name of pariah state] had nuclear weapons?' In the most obvious example, Iraq would have been in a position to move to regional pre-eminence through a combination of crude threats reinforced by occasional take-overs, while the West would have been inclined to stay clear, through fear of becoming involved in nuclear engagements. The fear of Iraqi chemical weapons reinforced opponents of military action in the run-up to the 1991 Gulf War: imagine the clamour if US and UK troops had been presented as prospective targets for nuclear weapons, or if Riyadh and Tel Aviv had been promised incineration. Kuwait would have been left to its fate. Whether or not an Iraqi nuclear capability could have been translated so readily into regional influence will be addressed later. For the moment, however, the key point is the redefinition of the proliferation problem as complicating the development of orderly regional politics in the move out of the strong bipolarity of the Cold War.

However, I have argued in this chapter that the bipolarity in central strategic relations went hand-in-hand in the past with disorder in a number of regions, so that the new situation is largely an extension of this tendency into Europe. The natural response by the major powers to this tendency is strategic disengagement. The case for active engagement has become increasingly problematic. It has come to reflect less a strategic imperative and more either a sense of *noblesse oblige* or more likely an awareness that economic and political chaos in significant parts of the world will have deleterious consequences.

Active engagement in these circumstances is unlikely to take the form of extension of alliances, unless the existing allies are convinced that new members will be relatively undemanding in terms of their security needs. This is despite the clamour for membership in the Western Alliance, from states who see such membership as a one-way bet now that there is no rival superpower to object. Western engagement is in fact more likely to take the form of economic-assistance packages, peace-keeping forces or conciliation services. More direct involvement may be confined, either to states where a security commitment has been in existence for some time, or in circumstances when (as with Kuwait) there has been a blatant act of aggression.

The world is approaching its maximum in terms of numbers of individual sovereign states. Many of these are chronically insecure,

yet few can expect their insecurity to be eased through membership of a formal alliance. Indeed alliance formation may now
become quite rare. An insecure state will have a natural interest
in a nuclear capability. This may not be its only option. For
economic or technical reasons, it may not even be an option. Any
attempt at acquisition may aggravate insecurity in the short term.
So, all that can be said is that the incentives for proliferation are
increasing. The reality may be far more reassuring.

If it took hold, proliferation would complicate casual intervention by Western powers in regional disputes, and this helps to
provide the current strategic basis for anti-proliferation policy. In
addition, of course, any use of nuclear weapons would not be
localized in its effects: fall-out does not respect national borders or
political sympathies. An active non-proliferation policy is, therefore, a natural aspect of the new world order. It is, however,
largely directed against new proliferation—which is why it is
not directed at south Asia and why, in the Middle East, it is
compromised by Israel's 'bomb in the basement'. It is easy enough
when directed against pariahs—such as Iraq or Libya. The problems will come if states which are recognized to be playing important roles in regional balances, or whom it is considered
undesirable to offend for pragmatic reasons, start to come under
suspicion. There could then be a tension between the commitment
to a global non-proliferation policy and the demands of regional
crisis management. All one can be sure of is that, once a grip is
lost on non-proliferation efforts, then crisis management in the
new world order will become progressively more difficult.

# 8

# Middle Eastern Stability and the Proliferation of Weapons of Mass Destruction

## YEZID SAYIGH

'THE Middle East has entered the nuclear age.' This was the terse summing-up by Israeli Defence Minister Moshe Arens towards the end of October 1991, as he surveyed the region's strategic environment in the aftermath of the Gulf War.[1] His remark was especially significant because it came only days before the opening session in Madrid of the Arab–Israeli peace process, a process supposed to lead, among other things, to a stable military balance between Israel and the Arab states and to regional arms controls in the conventional and non-conventional spheres.

Arens may merely have been reflecting on a reality that needs to be addressed (and presumably reversed) in the interest of regional peace and stability. Alternatively, he may have been staking out a pre-emptive position well in advance of any demands for arms controls and territorial concessions, by seeking to make the nuclear dimension (and implicitly the Israeli nuclear-weapons monopoly) an explicit and irreversible component of the regional strategic equation. Between the two interpretations lies the key to the impact of non-conventional-weapons proliferation on Middle Eastern stability.

The primary argument of this chapter is that weapons of mass destruction are inherently destabilizing in the Middle Eastern context, especially in the absence of an Arab–Israeli peace settlement and of a wider regional framework for the management of political, economic, and security issues. This is due, in particular, to the complex nature of the Middle Eastern environment and the

---

[1] Cited by former Defence Minister Ezer Weizman, *Yediot Aharonot*, 25 Oct. 1991.

high degree of interrelatedness between its various components, making stable balance, conflict prevention, and crisis management all the more difficult to achieve. Geographical proximity and multiplicity of regional actors mean that the other dimensions of stability and security cannot be isolated from the impact of non-conventional proliferation and the non-conventional arms race, nor can the strategic ramifications be confined to any pair or group of states.

For these reasons—small spaces and populations, proximity, inter-dependence—the presence of weapons of mass destruction in the Middle Eastern context is especially dangerous and de-stabilizing, and their use would be highly irrational.[2] Yet, para-doxically, the very same reasons appear to have led local decision-makers to the conclusion that non-conventional weapons might be useful as more than the ultimate strategic deterrent. The notion that they might also have battlefield uses—even in wars that do not threaten national survival—implicitly encourages the further conclusion that non-conventional conflict is wageable. This has already led to subtle shifts in the policies and doctrines of certain Middle Eastern states, and may ultimately give rise to the percep-tion that the possession of weapons of mass destruction confers the power of compellence, rather than solely of deterrence.

To break the pattern and prevent the emergence of a new, unstable balance of terror, or at least of non-conventional black-mail in the Middle East, any arms-controls proposals must take regional complexities and linkages into account. However, the linkages need not always be negative ones, that obstruct progress. Rather, deliberate use can and should be made of the interrela-tionships in order to effect trade-offs between the different dimensions and levels of regional security and stability. In this way, breakthroughs in arms controls and political settlements can be achieved that might otherwise be impossible or seriously delayed.

In the following sections, this chapter discusses in greater detail the four themes that determine the impact of non-conventional

[2] This is argued strongly by F. Barnaby, *The Invisible Bomb: The Nuclear Arms Race in the Middle East* (London: I. B. Tauris, 1989), 64–9. A similar argument is made in 'Establishment of a Nuclear-Weapon-Free Zone in the Region of the Middle East', *Report of the Secretary-General*, UN General Assembly, Document A/45/435, 10 Oct. 1990, p. 27 (hereafter referred to as UN Experts' Report).

proliferation on Middle Eastern stability: the nature and consequences of complexity; elements and politics of instability; implications for doctrine, and the constructive use of linkages to achieve arms controls.

### COMPLEXITY AS FUNDAMENTAL FEATURE

The Middle East is defined for the purposes of this discussion as a 'strategic system' that includes the Arab states and several non-Arab countries—Israel, Iran, and Turkey (some might add Ethiopia).[3] It is true that it may be divided into secondary 'security complexes' (or 'sub-complexes')—such as the Gulf, the Arab–Israeli theatre, the Nile Valley–Horn of Africa, and the Maghreb.[4] However, for reasons of geography, history, society, and culture, the degree of political permeability and strategic impact across the (notional) boundaries between these zones is high. It is more useful, therefore, to think of the region as a single, broad system, albeit a loosely structured one.

The multiplicity and diversity of regional actors render the Middle Eastern strategic system highly vulnerable to external influences, which, in turn, tend to reinforce the inherent fluidity of its internal relations. This is especially so given the relative proximity of southern Europe, the former Soviet Union, and south Asia (Pakistan and India), and, additionally, the global role and Cold War heritage that introduced the United States as a regional actor. Above all, therefore, the system is dominated by balance-of-power politics, and is thus characterized by its tendency to anarchic and violent relations.[5]

The above points to the *horizontal* complexity that is a basic

---

[3] The notion of the Arab states as a strategic system is developed in P. Noble, 'The Arab State System: Pressures, Constraints and Opportunities', in B. Korany and A. Dessouki (eds.), *The Foreign Policies of Arab States* (rev. edn.; Boulder, Colo.; Westview Press, 1991).

[4] The notion of the 'security complex' is developed by Barry Buzan in 'The Future of the South Asian Security Complex', in B. Buzan and G. Rizvi, *et al.*, *South Asian Insecurity and the Great Powers* (London: Macmillan, 1986), 235–52.

[5] There is a substantial body of literature on the Arab regional system, but, for recent surveys, see Y. Sayigh, 'The Gulf Crisis: Why the Arab Regional Order Failed', *International Affairs*, 67/3 (July 1991), 487–507; and R. Brynen and P. Noble, 'The Gulf Crisis and the Arab State System: A New Regional Order?', *Arab Studies Quarterly*, 13/1–2 (winter–spring 1991).

feature of the Middle Eastern strategic system. Of equal importance is its *vertical* complexity. This refers to the gradation of non-
conventional weapons and to those categories of conventional
military technology that have a special impact on the strategic
balance, such as ballistic missiles, anti-ballistic-missile missiles,
and certain advanced conventional munitions. Furthermore, there
is a distinct gradation of capabilities even within the conventional-
weapons category, while indigenous research and development
and production capabilities add a further dimension of complexity. The proliferation of the various non-conventional-weapons
categories is interrelated, and is further tied to developments in
the conventional weapons and political spheres.[6]

What makes the Middle Eastern strategic system particularly
unstable is the combination of its horizontal and vertical complexity. Despite the enormous expanse of some countries, the
critical 'conflict areas' are relatively small and tend to comprise
the borders of several states. As important, the vital concentra-
tions of population, and therefore of administration, economic
activity, and infrastructure, in rival states are often in close
proximity. Indeed, the distances separating the main regional sub-
complexes (with the possible exception of the Maghreb) are not
so great. With the proliferation of intermediate-range ballistic
missiles and mid-air refuelling for combat aircraft, most of the
major military powers in the region possess the means to threaten
the other key states, even those considered traditionally to lie in
other sub-complexes. The advent of non-conventional weaponry
and long-range delivery means may have given individual states
greater strategic reach, but by the same token it has also reinforced
linkages across the Middle East.

Factors of geography and technology dictate that the strategic
impact of non-conventional-weapons proliferation cannot be
contained within a specific sub-complex, let alone within a particular pair of states. A primary example of this is Iraq, which by
1990–1 was effectively extending its strategic reach into the Gulf
and Arab–Israeli theatres. It apparently started its nuclear-weapons
programme in response to the steep build-up of Iranian conventional and non-conventional capabilities under the Shah, but was
ultimately seen as a direct threat by Israel. Israel itself offers

[6] This is suggested in UN Experts' Report, 40–1.

another example, having demonstrated its reach both through ballistic missile and satellite launches (into the Mediterranean Sea opposite Benghazi, Libya, on one occasion) and by bombing targets as far apart as Baghdad and Tunis.

It may stretch the point somewhat (but not excessively) to anticipate the knock-on effect on Middle Eastern stability of Indian nuclear capability or the declaration in October 1991 by Pakistan that it had become a 'nuclear power'.[7] But it is easy to see that Iranian nuclear and chemical weapons and ballistic-missile programmes will affect the strategic posture and security not only of Iraq and the Gulf but also of Syria and Israel (and ultimately Egypt).

The strategic and, ironically, the political linkages between various sub-complexes and regional issues were further driven home by the US-led coalition effort against Iraq: first when the United States secured the active involvement of Egypt and Syria (and Arab states as far away as Morocco) and deployed allied forces to Turkey as well as the Gulf; and, secondly, when the United States moved in the post-war period to promote regional arms controls and achieve an Arab–Israeli peace settlement. Middle Eastern states have certainly built on these linkages in the past, with Israeli offers of nuclear-capable ballistic missiles to Iran in the 1970s and Iraqi–Egyptian collaboration on ballistic-missile development in the 1980s.[8]

Further destabilizing the Middle Eastern strategic system is the marked asymmetry of military capabilities in individual states. Vertical complexity has not been replicated across the horizontal spectrum of countries. A majority do not possess weapons of mass destruction and have little prospect of acquiring or developing them or their means of delivery, even assuming they have the interest in doing so (which is not the case for most). Of the minority of Middle Eastern states that can field non-conventional weapons, only Israel actually possesses all categories or the means to produce them, although Iraq was well on its way to acquiring similar capabilities before the 1991 Gulf War and the subsequent

[7] On Pakistani nuclear-weapons possession and policy, see al-Qods al-Arabi (London), 23 Oct. 1991; International Herald Tribune, 30 Dec. 1991; International Defense Review, 24 (Dec. 1991), 1306–7.

[8] The Israeli offer is cited in S. M. Hersh, The Samson Option (London: Faber & Faber, 1991), 274.

UN inspections. The probability that anti-tactical ballistic missiles and advanced conventional weapons and munitions will enter a growing number of arsenals in the region (or in its periphery) in the foreseeable future indicates that additional areas of asymmetry will arise.

At first glance, the asymmetry of non-conventional-weapons capabilities might not appear to be necessarily destabilizing. After all, 'excessive' proliferation of non-conventional weapons would only increase the danger of their use, whether by irrational design, misunderstanding, or accident. Furthermore, the emergence of a handful of 'hegemons' might establish a stable balance of power that would curb tendencies to open confrontation. Some might even argue that a monopoly of strategic power by Israel, nuclear-armed only for ultimate survival and under the restraining influence of the United States, could enhance stability, especially if it was accompanied by a peace settlement in which Israel could safely make the territorial concessions required by its Arab foes.[9] Another possible argument is to refer to the precedent of Cold War Europe, where the member states of the Western and Soviet alliances enjoyed peace for forty-five years, despite their large number and asymmetric capabilities (especially in the nuclear sphere).

In reality, the example of central Europe shows just why it is virtually inevitable that the proliferation or deployment of non-conventional weaponry will have a destabilizing impact in the Middle East. In the former case, the strategic balance was upheld between only two blocs, which between them accounted for almost all states in the northern hemisphere, with the minor exceptions of such neutrals as Austria, Sweden, and Yugoslavia. The United States and the Soviet Union were clearly in leadership of their respective alliances and underpinned that role with their massive nuclear arsenals and capability for global projection of power.

In contrast, the Middle East has a multitude of regional actors with divergent aims and policies, and lacks any semblance of stable balance. It suffers additionally from the active involvement of peripheral countries and out-of-area powers. Indeed, the only

---

[9] This is the central thrust of the argument by a number of Israelis, such as S. Feldman, *Israeli Nuclear Deterrence: A Strategy for the 1980s* (New York: Columbia University Press, 1982).

prospect of similarity with the central European experience is that of a new regional strategic balance, imposed by the unipolar projection of US influence in the 1990s and backed by a tenuous international consensus concerning the sanctity of internationally recognized borders. Paradoxically, though, if this prospect comes to pass, then it is likely to lead not to a new status quo based on weapons of mass destruction but to more active efforts to achieve non-conventional arms controls.

### CAUSES OF INSTABILITY

Inherent complexity and asymmetry may encourage instability in the Middle Eastern strategic system, but they are not its primary causes. Rather, these are the prevalence of conflict, arms races, and the nature of modern military technology. It is in that context that non-conventional weapons exert a strongly destabilizing influence, due to their own intrinsic features.

Conflict, arising from political or territorial disputes and exacerbated by social and economic factors, is the root cause of the regional race to acquire conventional and non-conventional weapons.[10] Examples abound: the Palestinian–Israeli and Arab–Israeli conflicts, Iran–Iraq rivalry and war, inter-state war and secessionism in the Horn of Africa, Algerian–Moroccan rivalry and the struggle over the western Sahara, the Libya–Chad border war, the Iraqi invasion of Kuwait and the 1991 Gulf War, and internal conflicts such as the Lebanese and Sudanese civil wars, armed Muslim opposition in Syria, and Kurdish and Shiite uprisings in Iraq. Further afield, confused security relations between the former Soviet republics and potential Turkish–Iranian rivalry in central Asia may impinge increasingly on the level of tension and instability in the Middle East.

In the examples cited above, both domestic and inter-state conflict has been violent, and has led to the militarization of society and the acquisition of large conventional arsenals. In several critical cases, the race to achieve strategic superiority has encouraged attempts by Middle Eastern countries to produce or otherwise obtain non-conventional weapons. The regional race

[10] The argument for the 'primacy of politics' in understanding and managing security is made in Y. Sayigh, 'Arab Regional Security: Between Mechanics and Politics', *RUSI Journal*, 136/2 (summer 1991), 38–46.

has in turn developed its own momentum, propelled by domestic, regional, and technical factors.

In the first case, efforts by specific regimes to remain in power, leadership perceptions of domestic prestige and regional status, and vested interests (commission-taking) have combined to fuel the continued drive for conventional and non-conventional capabilities. This is obvious in such distinct cases as Syria and Iraq or Kuwait, Qatar and the United Arab Emirates: there may be real defence needs, but the pattern of acquisitions reveals the strong influence of other priorities such as regime survival, the 'prestige factor' and the opportunity to earn commissions on arms contracts. To the foregoing should be added the conservative tendency of bureaucratic institutions to maintain existing policy and to protect jobs and budgets.

Secondly, the build-up of military capabilities in one country inevitably alarms its neighbours and alters the regional balance. This spurs efforts by the neighbours to 'catch up', and renews the sense of vulnerability (real or perceived) in the first country. Thus attempts to achieve national security through assured defence may ultimately undermine security and weaken defence instead.[11] The action–reaction build-up of weapons arsenals in the Gulf—Iraq, Iran, and the Gulf Co-operation Council states—since 1974 is a case in point. Conversely, the regional arms race may be driven by the attempts of some states to protect their refusal to resolve or concede political and territorial claims, the Arab–Israeli pattern of denial and counter-denial being a prime example. An added dimension is the tendency of the 'global' powers to perpetuate the regional status quo in its political and strategic dimensions, which puts certain local states at a disadvantage and drives their effort to compensate through non-conventional-arms acquisitions.

In the third instance, the arms race is driven by factors intrinsic to the nature of modern military technology. Cyclical obsolescence is a foremost example that compels modern armies constantly to acquire new generations of weaponry. Not only must each country prevent its military infrastructure and preparedness from dropping below a certain level, therefore; it has also to allow an additional margin for the lead-time between identification of new

---

[11] Point made in Abdul-Monem al-Mashat, *National Security in the Third World* (Boulder, Colo.: Westview, 1985), 13.

needs and technologies and subsequent acquisition of the requisite armaments.

The other side of the technological coin is that the constant appearance of state-of-the-art weapons systems and munitions—with increased lethality, range, and survivability—in the inventory of one army destabilizes the regional military balance and prompts countermoves. This is due to the magnified capability that such new technologies impart to their owners, even if overall force levels on the various sides of the regional balance are low to start with.[12] This is especially true of the emergent generation of advanced conventional weapons, which can be as devastating as non-conventional weapons if used against carefully selected targets, especially those necessary for the sustenance of life;[13] the same effect also applies, to some degree, to the advantages conferred by possessing in indigenous arms research-and-development and production capability.

## ENTER THE WEAPONS OF MASS DESTRUCTION

It is in the context described above that weapons of mass destruction are inherently destabilizing. In a regional environment marked by anarchic inter-state relations and balance of power politics, armed conflict, arms races, domestic struggles, and external intervention, the threat of mass destruction is especially worrying. In the continued absence of political solutions to deep-seated conflicts, acquisition of non-conventional arms by one state triggers counter-efforts by its rivals to do the same. What then makes the proliferation of non-conventional weapons especially destabilizing is their perceived utility for launching surprise attacks, demonstrative actions, and localized defence, as well as providing the ultimate deterrent.

Implied here is a crucial distinction between 'tactical' weapons and uses and strategic ones. True, this distinction may be one of perception rather than reality, and thus based on fundamental fallacies. It none the less encourages the view that non-conventional

[12] The power of advanced conventional weapons is discussed in various places in H. Rowen, *Intelligent Weapons: Implications for Offense and Defense* (Tel Aviv: Jaffee Center for Strategic Studies, 1989).

[13] A. Cordesman, *Weapons of Mass Destruction in the Middle East* (London: Brassey's, 1991), 167.

weapons can actually be employed in situations that fall short of a total war waged for absolute annihilation or absolute survival. In fact, with the obvious exception of nuclear weapons—and the less obvious exception of biological weapons, which in any case have not been fielded in the Middle East—the remaining categories of non-conventional and highly destructive advanced conventional weapons lend themselves quite readily to so-called tactical or operational uses. The same might even be seen to apply to tactical nuclear warheads, with high-yield weapons being set aside solely for strategic deterrence (the ultimate, 'last-resort' option).

The use of weapons of mass destruction of any description may be viewed as completely irrational in the Middle Eastern context, due to the proximity of rival population centres and the increased risk of mutual subjection to nuclear, biological, or chemical fall-out, but then it is no more rational in the central European or south Asian theatres. The notion of the tactical utility of non-conventional weapons might actually seem more attractive to Middle Eastern states facing relatively small foes along narrow fronts, than it did in the central European context during the Cold War, where the sheer size of armies, fronts, and populations and the certainty of massive escalation on a global scale acted as major disincentives against crossing the non-conventional threshold.

Even without actual use, weapons of mass destruction are widely perceived in the Middle East as more than an abstract, strategic deterrent. That they are considered as relevant to the conduct of war at the operational level was made evident by Iraqi use of chemical weapons against Iran during the first Gulf War, and of ballistic missiles against a nuclear-armed Israel in the second one. For Israel, which fears losing its conventional and technological superiority over the Arab armies, non-conventional weapons have offered a means since 1967 of constraining Arab war plans. These were kept limited and avoided encroaching on Israel's pre-1967 borders; the strategic limitation in turn influenced the choice of operational tactics. But it did not prevent war in October 1973, during which both Egypt and Syria were willing to resort to air-to-surface and surface-to-surface bombard-ment missiles when they considered that Israeli strategic bombing had exceeded certain red lines.

More recently, Arab anxiety has grown that the Israeli 'defensive shield' provided by non-conventional weapons can now be used assertively, as a strategic cover for conventional operations. More to the point, the fear is that, with its extensive first- and second-strike nuclear capability, long-range delivery systems, and evolving reconnaissance assets (including satellites), Israel is even in a position to wage an offensive non-conventional war yet remain relatively immune from counter-attack. Reinforcing this view is the Israeli effort to acquire or develop anti-tactical ballistic missiles, which are seen as completing the defensive shield and enabling Israel to make the aforementioned transition in posture and strategy.[14]

### DOCTRINAL SHIFTS

Arab fears are not entirely far-fetched. Although the specific scenarios mentioned above may not occur and Israel might not adopt a more offensive non-conventional posture, there has been a distinct blurring of conceptual boundaries concerning the function of non-conventional weapons and the doctrines governing their use. This applies in both the wider Middle Eastern context, where increasing ambiguity has led to the battlefield uses previously cited, and Israel, where subtle outward shifts in doctrine and policy have been underlain by far more concrete changes in force structure and deployment.

The negative implications of this trend for future regional stability are strong and clear. On the one hand, there is the constant danger of a shift in non-conventional-weapons policies at the strategic level, from deterrent to compellent thinking. On the other hand, there is a parallel shift towards incorporation of non-conventional capabilities in force structures and operational, war-waging doctrines.

Doctrinal shifts are most evident in the case of Israel, and specifically in relation to its nuclear posture. This became apparent as long ago as the October 1973 Arab–Israeli War, when Israeli nuclear-tipped missiles were put on the alert and deployed in their

---

[14] An Egyptian military analyst and retired general places anti-tactical ballistic missiles within the non-conventional category in the Middle East context (Ahmed Abdel Halim, 'The Un-Conventional Arms Race and the role of Arms Control Talks in Reducing the Risk of War', talk given at the Washington Institute for Near East Policy, reproduced by the National Center for Middle East Studies, Sept. 1990, p. 5.

firing positions. The Arab adversaries lacked the means to observe this move, which was intended instead to prompt the United States to accelerate its aerial resupply effort and, possibly, to persuade the Soviet Union to exert pressure on its Arab allies and limit their offensive. The nuclear deterrent was in fact used as a compellent. A similar demonstrative effect was employed following the Iraqi Scud attacks during the 1991 Gulf War, apparently to exert leverage on US conduct of its anti-Scud hunt.[15]

The compellent aspect of Israeli nuclear power, or at least the self-confidence and assertiveness it imparts to Israeli conventional force projection and inter-state relations, is evident in other ways too. Most prominent, and disturbing, has been the implied threat to the Soviet Union, starting with the southern republics and finally encompassing Moscow itself. The range achieved by Israeli ballistic missiles in test launches, coupled with the space programme and efforts (overt and covert) to obtain US intelligence data, indicated to both superpowers that Israel was seeking actively to target the Soviet Union.[16] This may partly have been an extension of the last-resort option, and a particularly dangerous one at that, but it was also an example of the compellence or leverage that Israel sought to exert in relations with the United States, by wielding nuclear weapons.[17]

It is reasonable to infer, moreover, that knowledge of its own non-conventional power has strongly reinforced certain directions in the strategic thinking of the Israeli leadership. Former Defence Minister Ariel Sharon's grand design of the early 1980s, in which he saw Israel's area of strategic interest as extending throughout the Arab world and into south Asia, is a case in point. The argument has been made that Israeli non-conventional-weapons capability can be a stabilizing factor, since it deters attack, and thus may allow even greater diplomatic flexibility. But it may also have underpinned past refusal to contemplate certain political

[15] According to various sources, Israel test-fired a nuclear-capable missile into the Mediterranean during the war, a signal of its readiness to use nuclear weapons against Iraq (*Jerusalem Post*, 10 June 1991).

[16] Hersh, *Samson Option*, 287, 290.

[17] This is clearly alluded to in Cordesman, *Weapons of Mass Destruction*, 173. Ironically some US officials and lobbyists were arguing that the United States should not raise the nuclear issue with Israel, so as not to lose US leverage over that ally (J. Nolan, *Trappings of Power: Ballistic Missiles in the Third World* (Washington DC: Brookings Institution, 1991), 100).

and territorial concessions, and then encouraged a manipulative approach to regional politics. Out of strategic superiority come notions of political domination.

Despite the dangers it poses, Israeli nuclear power has become widely accepted as part of Middle Eastern strategic reality. Whether due to the habituating effect of constant leaks or to the official US policy of 'no policy' on the matter, Israeli non-conventional weapons and means of delivery are regarded tacitly as part of the regional background. There was no Western reaction when Israel once again rejected an IAEA request to inspect its Dimona reactor in September 1991.[18] The following month, the US Administration decided to waive the financial sanctions that would normally be imposed on foreign-aid recipients, when the CIA revealed that Israel had been supplying South Africa with ballistic-missile parts and technology.[19] Nor was there public reference to the Israeli admission in December that half the heavy water imported from Norway twenty years earlier had been 'lost', suggesting possible diversion to its weapons programme.[20]

Complacency about the Israeli nuclear effort carries its own dangers, but more pertinent still is the extent to which nuclear weapons have apparently entered Israeli force structure and operational thinking.[21] Reports of artillery battalions being equipped with nuclear shells for 155-mm howitzers (and possibly 203-mm guns) and nuclear mines planted on the Golan Heights may be the result of deliberate disinformation, but neither can they be dismissed out of hand. Technological momentum and institutional inertia, as well as deterrent thinking, might indeed have led Israel into producing tactical nuclear weapons. In any case, Israeli nuclear devices may no longer be a weapon of last resort.[22]

The objection may be made, with considerable justification, that

---

[18] *Al-Hayat* (London), 22 Sept. 1991.

[19] *International Herald Tribune*, 28 Oct. 1991.

[20] *Jane's Defence Weekly*, 14 Dec. 1991, p. 1131.

[21] One Israeli analyst argues that there is a level of conflict at which Israel should confront Arab conventional strength with non-conventional means; and that the Israeli order of battle should be adjusted accordingly (A. Yaniv, *Deterrence without the Bomb* (Lexington, Mass.: Lexington Books, 1987), 256).

[22] Hersh, *Samson Option*, 276. The subtle shift in perception of what level of conflict should trigger nuclear defence is evident in the suggestion by Avner Yaniv that there should be a significant lowering of the 'invisible threshold' for such use (*Deterrence without the Bomb*, 256).

it is next to meaningless to distinguish too firmly between tactical and strategic non-conventional weapons and ballistic missiles in the Middle Eastern context, where distances are so short between rival capitals. And the taboo on crossing the nuclear threshold applies as much to tactical warheads as to strategic ones. None the less, there is a clear tendency for technological developments and institutional inertia to develop their own momentum and assert themselves, and so exert increasing pressure on official policy and doctrine from within. Furthermore, other factors may reinforce this trend, including the pressure of financial constraints on conventional force build-up, operational difficulties in a saturated Arab–Israeli battlefield, and the development of Arab military and technological capabilities.[23]

The issue is not so much that Israeli leaders might eventually believe that a war fought with non-conventional weapons is winnable, though there is such a risk. Leaders deciding to employ nuclear warheads, for example, might not expect or want to destroy the enemy utterly. Rather, it is more likely that they would detonate a single weapon demonstratively, to halt an enemy offensive at an early stage and abort it, before it poses a threat to national survival. (This is one effect of nuclear testing, but in its absence the demonstration might be carried out instead under real conditions.) While some elements in the Israeli space and anti-tactical ballistic missiles programmes might be construed as primarily defensive in purpose, the effort also lends itself to an evolving doctrine of controlled non-conventional-weapons applications, backed by real-time satellite intelligence and anti-ballistic missile defences. In all cases the risk is one of 'tacticizing' strategy, to use Yehoshafat Harkabi's term;[24] when wars are waged as if they were battles, perceptions of specific weapons and their uses shift.

The trend need not be limited to Israel. To some degree, Iraqi use of chemical weapons against Iran was intended to have a demonstrative effect, at least initially, as were early launches of ballistic missiles against Iranian cities. The means proved insufficient to achieve the purpose in the event, but that did not shake

---

[23] Discussed in Y. Sayigh, 'al-Jidal an-Nawawi fi Isra'il: ad-Dawafi' wal-Qadaya (The Nuclear Debate in Israel: Incentives and Issues)', *Shu'un Filistiniyyah*, 189 (Dec. 1988).

[24] Y. Harkabi, *Israel's Fateful Decisions* (London: I. B. Tauris, 1988), 94.

Arab belief in the potential effectiveness of ballistic missiles, for example.

Ironically, the 1991 Gulf War has probably reinforced Arab interest in ballistic missiles and other stand-off weapons, and the precedent of Scuds landing on Israeli cities has left a deep impression on Arab decision-makers and strategists everywhere (even in Egypt). They may see more accurate ballistic missiles, such as the North Korean Scud-P, as offering them a counter-force capability against purely military targets. The Syrian leadership, which is in the market for Korean Scuds and Chinese M-9s, might, for example, be encouraged to think that it can thereby deter Israel without triggering nuclear counter-use, because the Syrian threat would not be indiscriminate. Furthermore, the availability in open markets of navigation systems and electronics and the adaptability of certain areas of technology mean that developing countries may also be able to produce ballistic and cruise missiles with greater accuracy, range, and punch, and a lower vulnerability to anti-tactical ballistic missiles.[25]

The foregoing suggests that the various Middle Eastern actors will be increasingly attracted to what they perceive as the lower range of non-conventional weapons, that are so defined either because they are categorized as tactical (fallacious as the notion may be) or because they are sub-nuclear. The fact that they can still pack a punch in the restricted Middle Eastern arena makes them useful for both deterrent and blackmailing purposes.

As the red line on non-conventional proliferation drops lower, moreover, it may reach a point at which advanced conventional munitions become part of the strategic equation. Their strategic impact was demonstrated dramatically during the 1991 Gulf War, when they knocked out the heart of the Iraqi administrative, economic, and military systems. These and other emerging technologies, such as energy-based weapons, enable local states to wield considerable deterrent and compellent power, especially in the confined Middle Eastern context. Conversely, possession of advanced conventional systems by one state, such as Israel, may prompt greater Arab efforts to acquire nuclear capabilities as a counterweight.

---

[25] Nolan, *Trappings of Power*, 34; US Department of Defence experts, cited in *Jane's Defence Weekly*, 16 Nov. 1991.

In summary, two processes affect the evolution of weapons of mass destruction and their impact on Middle Eastern stability. One is the tacticization of strategy, while the other is the operationalization of its means. The advent of new technologies, such as advanced conventional munitions, is likely to accelerate both trends and make proliferation and actual use more likely, particularly in the absence of non-conventional-weapons and ballistic-missile capping and disarmament.

### THE POLITICS OF FUTURE INSTABILITY

The current strategic balance in the Middle East is very much in transition, as the region seeks a new equilibrium. Balance and equilibrium are quite distinct, and the regional trend towards non-conventional-weapons proliferation is largely a function of the degree of convergence or divergence between the two. Contrary to the notions of stability and quasi-permanence implied by the term 'balance', the Middle Eastern strategic equilibrium has been in constant flux over the past three decades. After the triggering of the Arab–Israeli conventional arms race in the mid-1950s, the launch of the nuclear and missile programmes in Israel and the chemical and missile ones in Egypt a few years later added a new, strategic dimension to the regional balance. That dimension has become part and parcel of disequilibrium ever since.

The massive military build-up in the Middle East (and the Gulf) that was prompted by the 1973 Arab–Israeli War and fuelled by the subsequent boom in oil revenues and external assistance contributed to the search for strategic advantage. In expanding their conventional power over the next few years, however, the main regional actors started to reach the limits of their strategic, technological, and financial capabilities. This prompted increasing interest in non-conventional arms, as suggested, for example, by Israeli nuclear co-operation with South Africa and the beginnings of the Iraqi nuclear-weapons programme.

The state of flux in the Middle Eastern strategic equilibrium became especially acute and prompted intensive counter-moves on all sides from the later 1970s onwards, following the withdrawal of Egypt from the Arab–Israeli conflict and the rise and collapse of the Iranian Shah's military ambitions. The consequences were not long in coming: the start of the Iran–Iraq War, the bombing of the Osiraq reactor, the Israeli invasion of Lebanon, and the Syrian acquisition of SS-21 missiles and striving for strategic parity, among others. The acceleration of the Middle Eastern non-conventional arms race in the second half of the 1980s, reaching its crescendo in spring 1990 with the Iraqi–Israeli exchange of threats and the invasion of Kuwait, was a logical and inevitable culmination of this growing trend.

These, then, are the dynamics that operate in the Middle Eastern strategic balance and that govern the direction and pace of change within it. The question now, in the aftermath of the 1991 Gulf War, is to determine how and why non-conventional weapons will continue to have a destabilizing impact on the region.

At present, and for as long as international sanctions and controls are in operation, Iraq has been neutralized as a regional contender with long-range non-conventional capability. It may be within the power of the global powers (the United States, first and foremost) to prevent Iraqi rearmament in the foreseeable future, but for how long can the resultant strategic imbalance in the Gulf be maintained without producing instability? Iran is apparently developing its own nuclear, chemical, and ballistic-missile pro-grammes, having restarted them at various times since 1984, and so will pose a growing threat to its neighbours in coming years. They need not respond directly as long as external powers uphold the existing strategic balance, but that option carries its own risks. Not least is the risk that suspension of the balance can increase disequilibrium, and can make disengagement by the external powers at a later date that much more problematic.

Two additional, recent developments are a cause for particular concern in this context. The emergence of Pakistan as a nuclear power has immediate implications for stability and security both in south Asia and in the Gulf. On the one hand, this affects the strategic balance with India and consequently with China, and so establishes the basis, however tenuous and nascent at first, for the convergence of the Middle Eastern and south Asian–Chinese security complexes.[26] Obvious examples are the negotiations in late 1991 over sale of an Indian nuclear reactor to Iran, which has

---

[26] This convergence is discussed in A. Ehteshami, *Nuclearisation of the Middle East* (London: Brassey's, 1989), 151–2, 154–6.

already received Chinese assistance in this field, and publication of US information concerning Chinese nuclear assistance to Algeria, Syria, and Pakistan.[27] This interlocking can only make non-conventional arms controls more difficult. On the other hand, the Pakistani nuclear capability may reinforce the emergence of an axis with Iran and Turkey. This has precedents, and closer co-operation has been discussed in recent years; any moves in this direction will certainly affect the strategic balance and perceptions in the Gulf and further afield in the Middle East.

The other recent development is the emergence of independent, nuclear-armed republics out of the chaos of the Soviet Union, of which one (Kazakhstan) is in central Asia. Further breakaways from the centre will produce a radically different situation along the southern borders of the former Soviet Union, with a range of consequences for Turkey and Iran and for the Middle East and south Asia beyond. One result will be to bring additional elements into the Middle Eastern strategic balance, making it far more difficult to achieve stability (let alone arms controls). This can occur even if the former Soviet Union plays no active part in regional politics.

The disadvantageous position of the Arab states might become more marked, especially in view of continuing Iranian efforts in the non-conventional-weapons sphere, while Israel may become even more resistant to nuclear disarmament. Having turned in the past to the Soviet Union for a protective strategic umbrella against Israeli nuclear power—an important factor in signing the Syrian–Soviet Defence Pact, for instance—Arab states might consider acquiring nuclear weapons to replace lost Soviet support.

The foregoing tends to confirm the view of the UN Experts' Report that, once states are in actual possession of non-conventional weapons, it becomes very difficult to establish a nuclear-weapons-free zone.[28] In the evolving Middle East, this means that two tiers of states are emerging: one with advanced non-conventional-weapons capability, and one without. Israel sits squarely in the former category and is likely to be joined, in the wider regional context, by Pakistan, Kazakhstan, and Iran. Some Arab states possess limited chemical weapons and ballistic missile programmes, but these remain modest in size and potential,

---

[27] *International Herald Tribune*, 16–17 Nov. 1991.
[28] UN Experts' Report, 26.

suggesting that the Arab states in general are set to occupy the second, non-conventional-weapon-free tier. With the MTCR in force and the CWC, this outcome seems increasingly likely.

Indeed, it seems unrealistic now to speak of Arab non-conventional-weapons capabilities. Iraq is steadily being shorn of all non-conventional-weapons power, Syrian ambitions are severely restricted—by lack of funds, the desire for better relations and trade with the West, and US pressure on China to cease missile exports—and Egypt is both strapped for money and extremely vulnerable to US economic leverage. Iran is a wild card at present, but Israel remains the single most important player, with its existing stockpile and advanced research-and-development and production capability.

This may, in turn, encourage smugness—the feeling that, at the end of the day, both the United States and Israel can, singly or jointly, 'manage' the situation and keep the Arab states from non-conventional-weapons proliferation. In effect, this would mean a continuation of the current US policy of no policy with regard to the Israeli nuclear programme. The Arab states are in dire need of cutting military expenditure, but they are unlikely to submit to the strategic imbalance indefinitely, especially as they will also consider it necessary to respond to developments in Iran and the southern Soviet republics. Conversely, should US policy start to shift, Israel might adopt the attitude that it is under siege from a hostile world. This, too, is dangerous, since Israeli nuclear monopoly cannot continue without prompting Arab countermoves.

Already, external constraints notwithstanding, the mixture of Israeli nuclear monopoly, Western neglect, and proliferation in the periphery of the Middle East has accelerated Arab nuclear efforts, as the cases of Iraq, Libya, Syria, and Algeria indicate. Egypt has repeatedly called for regional non-conventional-weapons disarmament to include all Israeli weapons of mass destruction, and in spring 1991 expressed its concern at proposals for Soviet–Israeli nuclear co-operation that might include sale of a 500-megawatt reactor to Israel for water desalination.[29] Its own effort to acquire nuclear power for civilian use has lain dormant since the mid-1970s, but its concern about Israeli capability continues to drive its non-conventional research-and-development programme. Even

---

[29] See, e.g., statement by U. al-Baz, *al-Hayat*, 7 Apr. 1991; editorial in *al-Ahram*, 30 Apr. 1991.

Saudi Arabia is rumoured to be interested in employing former Soviet nuclear scientists and may have sought to equip its Chinese-supplied CSS-2 missiles with nuclear warheads, reflecting its disquiet about Iranian efforts and the break-up of Soviet central Asia.[30]

So where does this leave the starting notions of balance and equilibrium? The Middle Eastern strategic balance will always be put under strain by new developments, both political and technological, that lead to shifts in the military and strategic capabilities of local states. Normally this would trigger countermoves, as other states seek to adjust and restore the balance. In this way equilibrium is maintained. However, attempts to prevent countermoves and so suspend the balance in its new configuration lead to disequilibrium and to deepening tension. This is particularly the case if non-conventional weapons enter the balance, and even more so if their introduction in one country is not matched in some way by a matching capability elsewhere.

### IMPLICATIONS FOR POLICY: USING THE LINKAGES CONSTRUCTIVELY

It is obvious that the complexity of the Middle Eastern strategic system reinforces inherent causes of instability. With the region in a state of transition at all levels—strategic, political and economic, external and domestic, and ultimately structural—the likelihood of further destabilization and conflict is enhanced. The view that the proliferation of non-conventional weapons might actually produce greater stability based on the 'balance of terror' is erroneous and dangerous.[31]

A more rational and far-sighted response is to recognize the wider linkages throughout the Middle Eastern strategic system, all of which meet ultimately at the political level. In the specific case of non-conventional weapons, the core link is the convergence of geography with the strategic and technological dimensions; the basis for moderating the impact of proliferation, containing and

---

[30] *Foreign Report*, Economist Intelligence Unit, 2142, Oct. 1991.

[31] Though not actually endorsing this view, Gen. Abdel Halim suggests that one option for discussion is for the Middle Eastern states to possess a specific number of non-conventional weapons by common permission ('Unconventional Arms Race' 5).

reversing it, is to address and resolve the political causes of conflict and militarization. The most practical course to achieve regional arms control, therefore, would be to embark on an active policy that deliberately utilizes the linkages that exist between the political and military and the non-conventional and conventional spheres.[32]

Building on linkages does not necessarily mean making movement in one dimension conditional on progress in all others, nor does it imply a particular sequence of resolution. The parties concerned should still seek, where possible, to promote negotiations on specific areas of arms control or political disputes independently of the other dimensions. Prospects for success in this endeavour will remain modest and vulnerable to reversal, however, unless the parties exert sustained efforts to resolve the other outstanding issues. It might prove impractical or counterproductive for them to propose well-defined visions of the future Middle East, but they should at least define policy guidelines and propose overall relationships between the various dimensions of stability and security.

Given progress along some tracks and convincing evidence of intent to move along the others, many states in the region would be happy to delink certain issues in order to cut spending and address other priorities. Accepting the linkages and utilizing them by the major powers would thus allow delinkage by the local ones. And, if major progress is made in resolving the central, political conflicts, then new dynamics could operate and alter the way in which the components of stability and security are perceived and managed. In the case of conventional and non-conventional arms controls, it would become easier to negotiate single issues, such as chemical weapons or ballistic missiles, if the parties acknowledged the relationship with other weapons of mass destruction and agreed in principle on the need for comprehensive and mutual disarmament.

In practical terms, all aspects of arms control must be addressed, both conventional and non-conventional, but this may be done in varying sequences. One option, more reassuring but less likely to happen, is for the various tracks to be pursued in parallel, along

---

[32] A particularly perceptive and coherent outline of the prerequisites for Middle Eastern arms control is in Cordesman, *Weapons of Mass Destruction*, 167–72.

with discussion of political issues. An alternative is to stagger negotiations and agreements, such that the separate tracks are dealt with consecutively; however, a vital condition for success is for the parties to express complete clarity and commitment regarding the subsequent stages of negotiations and their broad end results. This would apply to the order in which conventional and non-conventional arms controls are taken up, and also to separate items within each category.

As importantly, the alternatives of parallel or staggered negotiations would apply to the wider question of how to co-ordinate peace talks and regional arms controls, especially in the Arab–Israeli context. Controls are unlikely to be accepted by the local parties unless peace is secured, but mutual political and territorial concessions are unlikely unless military security is assured.[33] This suggests that specific proposals and timetables for conventional- and non-conventional-weapons reductions, and ultimately for disarmament and non-conventional-weapons-free zones, should be both pursued independently and presented as an integral part of negotiations for a political settlement.

Given an overall structure and timetable, it would become more possible to obtain asymmetric concessions at different times, since they would be matched later. In particular, the scale of Israeli conventional force reductions would be less than that of Arab neighbours, and nuclear disarmament and the establishment of a nuclear-weapons-free zone could be delayed until the conventional threat was removed. Concessions would be reciprocated in the interim, but in different kind, and the cumulative result would be a stable balance.

In formulating detailed proposals, special note should be taken of the fact that non-conventional weapons are far more destabilizing in a dense conventional-weapons environment. This is partly because the former might provide strategic cover for offensive operations, and partly because the threat of non-conventional use arises primarily in the context of large-scale conventional attack.

---

[33] This point is made at various points in G. Kemp, *Middle East Journal*, 45/3 (summer 1991). Conventional wisdom has it that creation of a nuclear-weapons-free zone should not be contingent on achieving a wider peace settlement, but in practice Israel insists on just such a linkage. For a general discussion, M. Karem, *A Nuclear-Weapon-Free Zone in the Middle East: Problems and Prospects* (New York: Greenwood Press, 1988).

In both cases, it is the war-waging capability of conventional weapons that presents the foremost risk, but the added threat is that non-conventional weapons are increasingly drawn into that capability.

Conversely, if conventional force levels are capped or reduced, then the justification for non-conventional weapons is reduced commensurately. Ultimately, the aim should be to bring conventional power down to a level that ensures balanced defence, while seeking to remove all non-conventional-weapons categories entirely, albeit in a staggered manner. Differences in capabilities and material circumstances mean different security needs and defence policies for individual states[34]—though this should not include possession of weapons of mass destruction—but all the more reason, then, for a multi-layered and multi-faceted concept of balance. Asymmetry can be an integral part of the final balance without causing disequilibrium or instability, so long as reciprocity and a high degree of mutuality apply.

Within such a balance, offensive capabilities would be constrained not only by force reductions, but also by deployment limitations and a basket of observation and verification measures.[35] These arrangements should also extend to indigenous research-and-development and production capabilities, and to the critical role of external suppliers of military technology.[36] Indeed, all arms-control agreements should be underwritten by clear limitations on supply, deployment, or other military intervention in the region by out-of-area powers.[37] A special forum would have the task of assessing and renegotiating the strategic balance in order to anticipate and accommodate the impact of emergent military technologies.

These and other aspects need to be addressed, and a variety of

[34] As argued in G. Kemp (with S. Stahl), *The Control of the Middle East Arms Race* (Carnegie Endowment for International Peace, 1991), 88.

[35] In addition to signing parallel treaties some measures are suggested by the UN Experts' Report, 42–3.

[36] This is particularly important because, despite the impressive indigenous research-and-development capabilities of countries like Israel, all Middle Eastern states remain highly dependent on technology transfer from the advanced industrialized countries. Point made in Nolan, *Trappings of Power* 40, and in R. Harkavy and S. Neuman, 'Israel', in J. Katz (ed.), *Arms Production in Developing Countries* (Lexington, Mass.: Lexington Books, 1984), 214–15.

[37] This is complicated by the diversity of suppliers, as noted in Nolan, *Trappings of Power*, 27.

detailed options must be offered. But the key argument is that
the mechanics of conflict resolution and arms controls can be
galvanized by restructuring and reordering the substantive issues,
allowing breakthroughs of a scale and scope previously considered
impossible.

## CONCLUSION

The proliferation of conventional and non-conventional weapons
in the Middle East has taken place against a background of
unresolved political disputes and active military conflict. This,
coupled with the prevalence of power relations in an anarchic
regional system and the absence of cohesive blocs of states, has
precluded the emergence of a stable strategic balance such as
existed for forty-five years in central Europe. Furthermore, the
instruments to bring the spread of non-conventional weapons
under control are still missing: the fact that the United States had
to bomb Iraq's non-conventional-weapons infrastructure in 1991,
ten years after the Osiraq bombing, revealed not only the inadequacy of the earlier Israeli attack but also that more effective
means of control had not yet been devised or used.

The last example is relevant because, in the aftermath of the
1991 Gulf War, there is a real opportunity to effect more sweeping
arms controls in the Middle East, albeit in an incremental or
staggered manner. Whether or not it will be seized successfully
depends entirely on the ability of the international community—
above all of the United States, with the backing of its leading
partners—to consolidate the dramatic change in the strategic
balance with substantial progress on the political issues that affect
the region at the inter-state and domestic levels.

Movement on the political issues is crucial, moreover, because
of the need to anticipate renegotiation of the NPT in 1995, and to
finalize, modify, or otherwise implement the CWC and BWTC
before then. Discussion of proposals for the establishment of a
nuclear-weapons-free zone in the Middle East should be revived
and given new impetus, especially as the convening of the regional
peace conference in October 1991 has opened serious prospects
for negotiating a nuclear-weapons-free zone directly between
Israel and the Arab states. Separate treaties governing the control
of specific non-conventional weapons could be negotiated in

parallel, and even mainly symbolic acts like agreeing test curbs would be useful as confidence-building measures.

The United States has a particular role to play in this context, though to do so it will have to revise its standing policy of ignoring Israeli nuclear capability. The United Nations would provide the most appropriate umbrella for such initiatives, while a Middle Eastern version of the Conference on Security and Co-operation in Europe could be the framework in which the various political, military, and economic dimensions are brought together, and in which the needs and circumstances of different states and sub-complexes can be articulated and balanced in a co-ordinated manner.

In conclusion, failure to achieve a formal Arab–Israeli peace, with concomitant non-conventional disarmament, represents the worst-case scenario for the prevention of non-conventional-weapons proliferation. In such circumstances, Arab counter-efforts are then inevitable, however piecemeal or delayed inviting further escalation and tension and leading to armed conflict or external intervention. Nor can the Arab–Israeli dimension be taken in isolation. A similar, politically oriented approach must also be applied in the other sub-regions of the Middle East, but in the Gulf especially, lest Iranian or other non-conventional efforts trigger Arab countermoves and thus destabilize the Arab–Israeli balance once more. The other linkage that must be controlled and utilized is between non-conventional and conventional power, with the need for transfer restraints, force reductions, and dis-armament in the latter area.

Middle Eastern states bear a primary responsibility in resolving regional problems, but so do their external supporters and suppliers. Ultimately, the international community as a whole has a role to play and a price to pay when it comes to non-conventional-weapons proliferation.

# 9

# Non-Conventional Weaponry and the Future of Arab–Israeli Deterrence

## AVNER YANIV

THE term 'deterrence' made its appearance in strategic discourse in the Middle East only in the 1960s. Until then, it had been used occasionally in political and strategic contexts but never in a technical sense and always without a specific meaning. The most important reason for the belated discovery of this notion—which, in Raymond Aron's words, is 'as old as humanity'—revolved around its initial ambiguity. At the beginning, it was far from clear what deterrence meant, and, even when it began to assume a more rigorous existence in the West, its relevance to what was happening in the Middle East was hardly self-evident.[1]

The Israelis were apparently the first to follow Western thinking in this regard. Some time in the late 1950s or early 1960s, the Israeli General Staff began to employ as a standard formulation the notion that the purpose of Israel's national security policy was deterrence. A leading member of the team of generals who turned this formulation into a virtual reflex was Yitzhak Rabin, who, in his tour of duty as Chief-of-Staff (1964–8), had the notion of deterrence embraced to such a degree that the late Dan Horowitz, one of Israel's foremost students of this topic, once declared it 'the most commonly used [term] of the jargon of strategic studies' and 'an integral part of the vocabulary of the public debate' in Israel.[2]

The Arab world was at first very slow to follow suit. The reason, at least in relation to Israel, was very simple. Deterrence, though supportive of offensive military means, is essentially a status quo strategy, clearly distinguishable from an aggressive

---

[1] R. Aron, *Peace and War* (New York: Doubleday, 1966), 404.

[2] D. Horowitz, 'The Israeli Concept of National Security and the Prospects of Peace in the Middle East', in G. Sheffer (ed.), *Dynamics of a Conflict* (Atlantic Highlands, NJ: Humanities Press, 1975), 244.

national posture in its heavy emphasis on the preservation of an existing international and/or regional order as its overarching political purpose.[3] Accordingly, as long as the Arab world was openly opposed to the continued existence of a Jewish state, it could not possibly adopt deterrence as its overall strategy, at least not *vis-à-vis* Israel. Indeed, what made deterrence so appealing to the Israelis was precisely that which made it so alien to their Arab adversaries.

This state of mind began to change in the wake of the 1973 Arab–Israeli War. Egypt and Syria launched their offensive at a time of their own choosing and succeeded in taking Israel by surprise. None the less, after less than three weeks of fighting, the IDF was on the verge of yet another decisive military victory. To escape this calamity Egypt had to endorse accommodation with Israel as its explicit goal, a change which in one fell swoop turned it from a revisionist to a status quo power—namely, precisely that kind of an international actor for which deterrence is a natural choice.

Even before the Egyptian–Israeli peace process was consummated, the Arab world as a whole adopted deterrence as the explicit purpose of the (mainly Syrian) force which was deployed in Lebanon in order to put an end to the civil war.[4] Then came Camp David, on the one hand, and the disastrous Iraqi invasion of Iran on the other. Syria was left all by itself to face Israel under the Likud, which, on top of that, had doubled its military strength since the 1973 Arab–Israeli War. Under these alarming—from the Syrian point of view—circumstances, strategic parity, a somewhat muffled title for what in effect was a full-blooded deterrent posture became the be all and end all of Syrian national security policy. In the words of Asad's biographer, 'Asad's Syria represents the rejection of an Israeli-dominated Middle East order, offering instead one based on the supremacy of neither Arabs nor Israelis but on a balance of power between an Arab Levant centered on Damascus and an Israel within its 1948–9 bound-

---

[3] For a more detailed conceptual analysis of this topic, see A. Yaniv, *Deterrence without the Bomb: The Politics of Israeli Strategy* (Lexington, Mass.: D. C. Heath & Co., 1987), 5–12.

[4] For details, see N. J. Weinberger, *Syrian Intervention in Lebanon* (New York: Oxford University Press, 1986), 255–6, 325–6.

aries'.[5] At approximately the same time, there was a surge of articles and books calling upon Israel and the Arabs to adopt a nuclear posture. Such a move, some of these writers proposed, would accelerate a process whereby the Arabs would seek their own nuclear capability. In turn, one writer confidently predicted, 'a stable system of mutual nuclear deterrence' will emerge, and at last freeze this festering wound in place.[6]

Even though the advocates of this idea were very close friends of Israel or, indeed, Israelis themselves, the Israeli establishment neither shared their view nor encouraged them to propagate it. To the contrary, official policy continued to shun publicity on the nuclear issue and would go to almost any length, including an attack on the Iraqi nuclear installations and the kidnapping of the hapless Vanunu, in order to avoid an explicit nuclear posture.

However, if this was intended to slow down Arab efforts to develop a countervailing nuclear capability, it was only partially successful. Egypt, Israel's most formidable Arab adversary in the past, quietly acquiesced. Syria developed a significant chemical capability as a cheap substitute for a nuclear deterrent. Iraq, defying the 'Israeli *diktat*', pushed relentlessly ahead with a massive undercover effort to develop a variegated and exceptionally well-protected nuclear programme. Had it not been for Saddam's miscalculation on 2 August 1990, this would have earned Iraq a nuclear capability about now.[7]

Whether or not the Iraqi programme would have resulted in an enhanced mutual deterrence is essentially the key question which this chapter seeks to address. It is not a simple question, because of the secretive nature of the nuclear programme both in Israel

[5] P. Seale, *Asad: A Political Biography* (Berkeley, Calif.: University of California Press, 1988), 495.

[6] S. J. Rosen, 'A Stable System of Mutual Deterrence in the Middle East', *American Political Science Review*, 71 (1977), 1367−83. See also S. Aronson, *Israel's Nuclear Options* (Centre of Arms Control and International Security (ACIS), Working Paper, 17; Los Angeles: Centre of Arms Control and International Security, University of California, Nov. 1977); R. Tucker, 'Israel and the United States: From Dependence to Nuclear Weapons?', *Commentary*, 60 (Nov. 1975), 29−43; S. Feldman, *Israeli Nuclear Deterrence: A Strategy for the 1980s* (New York: Columbia University Press, 1982).

[7] For a detailed, up-to-date and well-informed survey of the non-conventional arsenals of Israel and its adversaries, see A. H. Cordesman, *Weapons of Mass Destruction in the Middle East* (London: Brassey's, 1991).

and in Iraq. What can be done at best is to speculate, using some published material and extrapolating for the balance from the experience gained in four decades of thinking about these issues in the somewhat different context of East–West relations.

## THE MILITARY BALANCE

Before Operation Desert Storm, any simple and straightforward capability analysis would have led to the conclusion that the days of Israel's 'bomb-in-the-basement' posture were numbered. The Jaffe Centre for Strategic Studies at Tel Aviv University, to quote an authoritative example, estimated that an 'Eastern Front' comprising Syria, Jordan, some Lebanese militias, and contingents from Kuwait, Libya, and Saudi Arabia would deploy 595,000 troops (440,000 regulars and 155,000 reservists) on land, 138,000 (105,000 regulars and 33,000 reservists) in the air, and 14,200 (13,000 regulars and 1,200 reservists) at sea. The IDF would deploy in such a war 540,000 soldiers (170,000 regulars and conscripts and 340,000 reservists). Such an Arab 'Eastern Front' would be able to count on the support of a far larger Arab coalition consisting of Egypt, Algeria, and Morocco—but not Iraq—and have at its disposal (according to Jaffe Centre estimates) 1,582,000 troops (of whom 920,000 would be regulars), or 3.6 times more troops than the IDF at its largest configuration. These Arab forces, in the assessment of Major-General (ret.) Aharon Yariv and his team of researchers (most of whom are veterans of the IDF's Military Intelligence Division), would be able to field 7 armoured and 6 mechanized divisions in a minimal 'Eastern Front' configuration, but 17 and 15 such divisions respectively in an all-Arab configuration. To this should be added 9 infantry divisions and 23 brigades of Paratroopers and Commandos in a larger configuration, compared with 2 infantry divisions and 15 paratrooper brigades in an 'Eastern Front' scenario. In terms of war material, this would mean 4,680 tanks in an 'Eastern Front' configuration but 7,990 in the larger alternative; 5,350 armoured personnel carriers, in a limited scenario and 10,050 in the more inclusive alternative; 2,616 artillery pieces in the limited configuration as opposed to 5,120 in the wider alternative; 3,450 surface-to-air missile batteries in a narrow configuration and 5,724 in the larger configuration; 86 surface-to-surface batteries in a limited configuration or 129 batteries in a larger configuration; 805 combat aircraft in a limited configuration or 1,305 in an all inclusive configuration, and, finally, 376 combat helicopters in a narrow 'Eastern Front' as opposed to an estimated 528 such helicopters in a more inclusive Arab coalition.[8]

Another estimate by Major-General Avraham Rotem, which excluded Iraqi participation, came up with even more staggering figures than those of the Jaffe Centre's team. Rotem spoke in a public lecture at the Jerusalem Van Leer Foundation:

In a wide canvas, it is possible to say that an Arab coalition consisting of Syria, Jordan, Iraq and Saudi Arabia, attacking from the east with a limited Iraqi participation but without Egyptian participation and with only token contingents from other Arab countries, will present Israel with a formidable force of 7,000 tanks, 1,500 quality armoured personnel carriers of the BMP variety, 200 combat helicopters, almost 3,000 artillery pieces, 50–60 surface-to-surface missile launchers (more than half of these are Scuds and SS-21s), about 1,000 advanced aircraft and altogether some 800,000 or even 900,000 officers and men. Egyptian participation can increase [Arab] land forces by 25 per cent and air forces by 30 per cent. This is a force three times as large as the force against which Israel fought in the Yom Kippur war, and it contains in real numbers qualitative elements which did not even exist during that war. . . . Israel, for its part, will be able to counter this force by fielding far greater forces than it could at the Yom Kippur War and these will have some special features of high quality. The estimated correlation of forces . . . will be . . . 4 to 1 in manpower, 3.5 to 1 in tanks, 7.6 to 1 in artillery pieces, 3 to 2.1 in division size combat units, 3 to 2.1 in assault helicopters, and 4:3 to 1 in combat helicopters. It is plausible to assume that from the Israeli point of view this quantity will exhaust the resources which were available, are available and will be available until this war.[9]

Desert Storm, however, led to significant revisions in the estimates. The damage to Iraqi infrastructure was immense. Iraq's external debt after the war was estimated at about $200 billion. The Iraqi army was not annihilated, but it suffered a major setback and, even ignoring the weakness of the Iraqi economic and physical infrastructure after the war, would need several years

[8] Jaffe Center for Strategic Studies, *Judea and Samaria: Alternative Roads to a Peace Settlement* (Tel Aviv: Tel Aviv University, 1989), 176–83.
[9] A. Rotem, 'The Next War', in A. Hareven (ed.), *Approaching the Year 2000: Towards Peace or Another War?* (Jerusalem: Van Leer Foundation, 1988), 128.

to regain its level of force prior to the war. Israel, conversely, obtained quantities of additional weapons—mainly relating to its anti-aircraft capability—as a direct result of the crisis and, especially, of the US effort to stop it from retaliating against Iraq's Scud missile attacks.[10] Syria, too, has used some $2 billion in hard currency, obtained from various sources (Saudi Arabia, the United States, Japan) as a means of ensuring its participation in the anti-Iraqi coalition, to acquire hefty quantities of additional weapons in both Eastern and Western Europe. The same applies to Turkey and, above all, to Saudi Arabia, which firmly rejected US requests for massive prepositioning of American arms on its territory but demanded a huge arms deal with Washington to beef up its ability to stand up to Iraq (or, for that matter, Iran) in the future. The total size of the Saudi deal appears to be $23 billion, and it will be implemented over most of the 1990s.[11] As a result of all this arming by Iraqi neighbours, the Iraqi conventional threat to Israel appears to have diminished to its level in the 1970s.

This estimate relates, of course, to the bilateral Israeli–Iraqi balance. But Israeli military planning cannot isolate the Iraqi issue from the military build-up of other Arab states. Iraq under a defiant and megalomaniac Saddam Hussein has had a galvanizing influence on all of Iraq's neighbours. In the context of the Gulf crisis *before* Desert Storm, Iraq focused on itself the attention of Syria, Turkey, Iran, the Gulf states, and Jordan, and thus reduced the menace which two or three of these powers could have posed to Israel in a different context. Yet, the diminution of the Iraqi threat immediately restores the more traditional Arab–Israeli military balance. If, for instance, the Madrid Conference and the peace process which it represents are bogged down, there is bound to be a rise in the anticipation of hostilities between Israel and the neighbouring Arab states, especially Syria. In turn, the hastened pace of arms procurement by Syria, Saudi Arabia, and even Egypt has generated proportional pressures for further arms procurement by Israel.

[10] See Z. Eitan, 'The Iraqi Threat to Israel after the Gulf War', in Jaffe Centre for Strategic Studies, *War in the Gulf: Implications for Israel* (Tel Aviv: Papyrus, 1991), 137–40.
[11] For details, see D. Gold, 'The Gulf Crisis and US–Israel Relations', in Jaffe Centre for Strategic Studies, *War in the Gulf*, 75–6.

## CRISIS STABILITY

The record of Arab—Israeli crises over a period of four-and-a-half decades is not as utterly depressing as meets the eye. It is true, of course, that the bitter feud between Arabs and Jews has produced numerous crises, some of which have led to enormous bloodletting. Yet, taking a longer-term perspective, it is arguable that the trend is towards an enhanced capacity for effective crisis management.[12]

The starting-point for such an evaluation should be the recognition that, notwithstanding the numerous crises in the 1950s, most of them were of a limited scale and hence insufficient to trigger a full-scale war. Indeed, Israeli Chief-of-Staff during 1953–7, Lieutenant General Moshe Dayan, who was convinced as of mid-1955 that war was in Israel's interest, could only hope to lead matters towards the crucible of war through an ever-accelerating sequence of incidents. None of these was sufficient, on its own, to bring about war; but over time he hoped that they would create a snowball effect that would lead to the showdown which he wished to create.

The first real Arab—Israeli crisis requiring crisis management to prevent war occurred in February 1960, when Egypt deployed five hundred tanks in the Sinai Peninsula, in order to divert Israeli attention to the region, while Syrian forces were being withdrawn from the Golan Heights in order to take part in a large-scale parade which Nasser, in his capacity as Head of the then Egyptian—Syrian Republic, was going to inspect in person in Damascus. Israel was caught—as Yitzhak Rabin wrote in a note to his colleague, Major General Ezer Weizman, during an emergency General Staff meeting—'with its pants down'. The IDF was put on alert, some reserves were called up, and the crisis was quietly and peacefully defused (presumably 'managed') within seventy-two hours.

This seeming success was on the minds of both Israeli Prime Minister Eshkol and Egypt's President Nasser in May 1967—except that both drew the wrong lessons, took the wrong steps, and thus reached a point at which war became the only way out. The result was the devastation of three Arab armies in June 1967, and a follow-up war of attrition between Israel, on the one hand,

[12] This section draws mainly on my forthcoming book *Politika veestrategya be-Israel* (*Politics and Strategy in Israel*) (Tel Aviv: Sifriyat Poalim, 1992).

and Egypt, Syria, and the PLO on the other hand, as well as a number of dangerous crises. The most striking of these occurred in late September 1970, when Syria sent an armoured column into Jordan to help the PLO, which had been ferociously pursued by King Hussein's troops. What helped manage this crisis was a combination of deliberate and fortuitous factors. The then Syrian Minister of Defence, Hafiz Asad, was cautious and would not go to war with Israel in order to save the PLO. The United States projected itself into the scene, which had a sobering effect on the key players. Finally, Israel was very cautious and would not play brinkmanship without iron-clad guarantees of backing from the United States, which President Nixon was not quite ready to offer.

The US intervention in the Jordan crisis foreshadowed to a certain degree the next crisis in Arab–Israeli relations—namely, the 1973 Arab–Israeli War. When the war broke out, there was no question of crisis management, since Egypt and Syria were determined to undertake this endeavour. Towards the end of the war, however, a classical crisis occurred when, after the 22 October ceasefire, Israeli forces continued their drive to encircle the Egyptian Third Army. During the next forty-eight hours the world held its breath while the United States and the Soviet Union engaged in an intricate manœuvre which ended with a second ceasefire. Thus, outside intervention both precipitated and contained this crisis virtually on the brink.

Three years later, yet another potentially explosive crisis was managed with a good deal of third-party (superpower) intervention. Following the near collapse of the Maronite showdown with the Lebanese Left and their PLO allies, Syria positioned itself to move into Lebanon in order to restore stability on its own terms. A long-standing Israeli policy suggested that Israel would object to such a massive projection of Syrian power inside Lebanon. The United States offered its good offices, and the result was the so-called 'red-lines' agreement. Israel accepted Syrian intervention in Lebanon within very specific limitations, whose purpose was to safeguard Israel's own interests in the Lebanese quagmire.

The red-lines agreement held for five years but then collapsed under the weight of the deteriorating crisis in Lebanon, which, in turn, drew both Israel and Syria into ever-growing involvement. In April 1981 a deliberate provocation by Bashir Gumail's 'Lebanese Forces' brought about a fierce Syrian reaction. Gumail then called on Israel for help, and a divided Israeli cabinet gave its grudging consent to an air strike against Syrian helicopters operating near the Lebanese town of Zahleh against a Lebanese Forces' position on Mount Senin. Two Syrian transport helicopters were shot down by the Israeli Air Force and Syria responded by introducing surface-to-air missiles into the area. Israel threatened to retaliate immediately if the missiles were not withdrawn. Fearing a conflagration leading to war, President Reagan sent a special envoy, Ambassador Philip Habib, to mediate. Habib failed to resolve the crisis, but there is little doubt that his presence in the region, signalling, as it did, the enormous importance which the United States attached to the prevention of another Arab–Israeli war, had a sobering impact on the rivals. Indeed, the crisis was defused subsequently, owing to fortuitous factors such as weather conditions. A year later, during the Lebanon War, the Israeli Air Force was ordered to strike at the Syrian missiles, and did so with utmost success.

Then, nine years later, the United States once again managed an Arab–Israeli crisis and prevented a fully blown war. The occasion was Desert Storm. Seeking to split the US-led coalition, Saddam Hussein ordered Scud attacks against Israeli cities. Israel was on record for an unqualified commitment to retaliate instantly and massively. Fearing an uncontrolled chain reaction following an Israeli retribution, the Americans embarked on a sustained effort to destroy the Iraqi Scud launchers, while at the same time applying massive pressure on Israel to continue its policy of restraint. Since the number of Israeli casualties from the Scud attacks remained negligible, Israel obliged. But by the end of the crisis it was clear that a larger damage to Israel would have triggered an active Israeli response. As on previous crises, then, the omnipresence of the United States had a moderating influence, but the destabilizing pressures generated by the wild dynamics of Arab–Israeli politics were still enormous. Arab–Israeli relations on the whole have clearly mellowed to a point at which crisis management has become a feasible proposition. But, as in other similar cases, the need for extensive crisis prevention through political settlements was also underscored with particular gravity.

## PERCEPTUAL PROBLEMS

One of the most stubborn fads about the sources of conflict in the Middle East is that it is all a reflection of a grand misunderstanding, whose origins are in the twisted minds of Arabs, Israelis, and other nationalities in the region. Aware of this, Michael Brecher placed what he calls 'the psychological environment' at the centre of his studies of Israeli foreign policy and crisis behaviour.[13] Similarly, Herbert Kelman has been advocating similar ideas for decades,[14] while Daniel Heratsveit has followed all this with surveys and interview material.[15] David Shipler has made a best seller of his moving descriptions of 'wounded spirits in the promised land'.[16] All of these writers—and many less successful ones—highlight a critical factor in Middle Eastern conflicts as well as in any conflict. But, arguably, all of them confuse secondary with primary factors. To say that there are deep misperceptions between Arab and Jew is to state the obvious. But to make it sound as if this is the only factor which really accounts for the depth of the conflict is to indulge in fantasies. The Arab–Israeli conflict does not start from misperceptions; quite the contrary. The conflict has been vicious, bloody, and intractable, because the contenders have known exactly all along what their opponents were up to. As Israel's founding father, David Ben-Gurion, once told the great Zionist leader Nahum Goldman,

Why should the Arabs make peace? If I were an Arab leader I would never accept the existence of Israel. This is only natural. We took their land. True, God promised it to us, but what does it matter to them? There was anti-Semitism, the Nazis, Hitler, Auschwitz, but was it their fault? They only see one thing: we came and took their land. Why should they accept it? They may forget in a generation or two, but for the time being there is no change.[17]

The same stress on the objective validity of the Arab commitment to the struggle against Israel, but in an even less expected way, is also central in the thinking of Ezer Weizman—a one-time

[13] M. Brecher, *The Foreign Policy System of Israel: Setting, Images, Process* (Oxford: Oxford University Press, 1972).

[14] H. C. Kelman, 'Israelis and Palestinians: Psychological Prerequisites for Mutual Acceptance', *International Security*, 3/1 (1978), 162–87.

[15] D. Heradsveit, *The Arab–Israeli Conflict: Psychological Obstacles to Peace* (Oslo: Universitetsforlaget, 1981).

[16] D. K. Shipler, *Arab and Jew: Wounded Spirits in the Promised Land* (New York: Times Books, 1986).

[17] N. Goldman, *The Jewish Paradox* (Jerusalem: Israel Universities Press, 1968), 99.

hawk turned into a leading dove. 'We [Israelis] have adopted the view', he wrote after the 1973 Arab—Israeli War,

that the Arabs are mystics, while our strength stems from our rationalism. But an objective consideration of the circumstances and of the numerical proportions reverses the picture: it's we who are the mystics and the Arabs who are the rational realists. We claim that despite everything three million Jews will withstand 100 million Arabs. They believe that, in the long run, their enormous numbers and their fabulous wealth will give them the advantage. To win, they reason, we [i.e. the Arabs] do not have to be as good as the Jews on the battlefield. It's quite sufficient to be a lot worse than they [the Jews] are for our quantities will overcome ... [Israel's] qualitative superiority. The Jews have tensed their muscles as far as they will go. Now we'll press forward—not to draw level with them, but just to reduce the gap between us. That will be enough to beat them and wipe our disgrace.[18]

Were Arab perceptions of Israel to be identical with those of the Israeli leaders quoted above, it would be possible to argue that no barrier of misperceptions separates the two sides at all. But, although the dissimilarity is not as glaring, or as central, as perception theorists would have it, it is nevertheless there and does have an important impact on Arab behaviour towards the Jewish state.

Oddly enough, both Israeli and Arab perceptions of the conflict are predicated on an inherent inner contradiction: the Israelis tend to view the conflict as a product of asymmetry in power between an Israeli 'David' and an Arab 'Goliath', which, as Weizman's words point out, breeds an asymmetry of objectives between Israel and an Arab world, the former seeking acceptance and accommodation while the latter seeks Israel's annihilation. Yet, at the same time, Israeli views of the ultimate future of the conflict are full of self-assurance, and stress Israel's ultimate ability to withstand the test and survive. The Arabs, for their part, at once emphasize the weak and transient nature of Israel, which, like the Crusaders' kingdom in the twelfth century, is bound to be merely a passing episode, and Israel's superior power and relentless drive for regional hegemony.

The point at which both warped and paranoid images meet is that neither Israelis nor Arabs deny that Israel is small and the

[18] E. Weizman, *On Eagles' Wings* (New York: Macmillan, 1976), 155.

Arab world large. In turn, the main divergence hinges on the interpretation of Israel's *ultima ratio*. For most Israelis, even when the IDF is unleashed in a preventive or pre-emptive war (as in 1956, 1967, and, to a lesser extent, 1982), it is acting defensively. For most Arabs, even when Israel is the victim of a carefully orchestrated attack (as in 1948, 1973, and, to a lesser degree, 1991), it remains the aggressor.[19]

These fundamental dispositions reflect directly on attitudes to any military build-up, and especially to the acquisition of non-conventional weapons. If Israel is inherently illegitimate and aggressive then its acquisition of non-conventional weapons can only mean—at least to Arab radicals—a brazen drive to regional hegemony. Once it has such weapons, it will—or so Arab leaders often affirm—use them in order to impose its will on its regional environment. The fact that Israel has had access to nuclear technology and sophisticated delivery systems for more than two decades now, but has not used this in order to bring its neighbours to their knees, does not mean that it will not do so in the future. The Israelis, as Arab commentators see it, are anxious to avoid friction with the United States. Therefore, they let the world get used to Israel's nuclear status gradually, over time. Eventually, when Washington and the rest of the West accept Israel's nuclear status as an irreversible *given*, the Israelis will flex their nuclear muscles.[20]

Against the background of the above discussion, to speak of 'arms-race stability' in the context of the Middle East is to employ a virtual contradiction in terms. Indeed, this is not a semantic issue but a grave reality. Any retrospective overview of the Arab–Israeli arms race during the four and a half decades since the establishment of the state of Israel will immediately underscore a fundamental dynamics which makes a stable equilibrium look like a pipe dream. This is in reference to three critical factors: the premisses of security policies of the main adversaries; inter-Arab rivalries; and Arab rivalries with non-Arabs in the outer ring of Middle Eastern states.

The most fundamental premiss of Israel's national security so far has been that, at any given moment, the Jewish state requires

---

[19] Most of what is said in this part of the discussion draws heavily on M. Klein and M. Steinberg, *The Arab World in the 1980s* (Jerusalem: Leonard Davis Institute, 1990).
[20] Ibid.

military force of sufficient firepower to be able to score a decisive victory against any combination of Arab forces, even if the latter act in concert and have the advantage of complete surprise. Such a working assumption has been based on the combined experience of four decades and half-a-dozen wars. What this conception has boiled down to has been a force level relating not to the military power of one Arab adversary but to the military power of all key Arab states combined.[21]

The trouble, however, has always been that, from the point of view of all key Arab players, it has been impossible to make the same assumption. What Egypt and Syria in particular have logically been impelled to assume has been that Israel could—and therefore, in their thinking, would—concentrate most of its formidable arsenal against one of them at a time. This was the Israeli strategy in 1948 when the Hagana irregular forces, and subsequently the newly formed IDF, offset its numerical inferiority by concentrating first against the Palestinians, then in quick succession against the Arab Legion, the Lebanese Army, the Syrian Army, and finally the Egyptian Army. The same strategy was perfected further in the 1956 Sinai campaign and in the Six-Day War (in which Israel launched a surprise attack against Egypt alone), and then implemented once again in 1973 (when the IDF at first contained the Egyptian army and launched a major offensive against the Syrians, and then moved to containment in the Golan while launching its offensive across the Suez Canal), and, of course, in 1982, when the PLO and Syria were taken on alone. It made ample sense for the Israelis to try and offset their quantitative inferiority by concentration of force and the creation of local superiority in key sectors of the war. But it also made ample strategic sense for Egypt and Syria to assume that they could be Israel's next prime target. The upshot in fact has been that a real equilibrium in the arms race has been inherently unattainable, since what would be adequate by Israeli definitions would be a menace from the Arab point of view, whereas what would be adequate from the Arab point of view would pose an existential threat from the Israeli point of view.[22]

---

[21] For an analysis of the mute debate in Israel over this issue, see A. Cohen, 'Israel and the Atom: The Debate That Never Was', in A. Cohen (ed.), *Humanity in the Shadow of the Bomb* (Tel Aviv: Hakibbutz Hameuchad, 1987), 69–82.

[22] S. Aronson, 'The Theory of Nuclear Deterrence and the Situation in the Middle East', in Cohen (ed.), *Humanity in the Shadow of the Bomb*, 231–2.

Inter-Arab fights, games, and debates have obviously made things worse. As a long-term process, the Arab world has been clearly moving away from Pan-Arabism to a far looser commitment for joint action. As Fouad Ajami has shown, Nasser brought the idea of literal Arab unity *ad absurdum*;[23] his death thus symbolized the passing away of a grand but hollow idea. Since then it has become legitimate for each and every Arab state to look after its own narrower interests. The result has been a mere acceptance of something which had been a reality all along: Israel's adversaries may speak the same language and broadly have a common history, but they constitute, nevertheless, a regular system of states in which the ground rules are derivative of the age-old balance of power.

The assumed power of the Hashemite kingdoms of Iraq and Jordan in the 1940s galvanized a loose coalition of the rest of the Arab world against them. The ascendance of Nasser in the 1950s brought about a major schism between 'radicals' and 'conservatives'. Syria's concern to offset Iraq's rising power in the 1980s brought about an alliance between the ultra-secularist Ba'th regime and the ultra-clerical Iran of the Ayatollahs. In most of these cases the military dimension was seldom overt. But it was always there and generated inexorably additional pressures to acquire arms. Thus, even though most of Syria's military might confront Israel, its arms purchases must also be influenced to a degree by a deep concern not to expose itself to a position of such military inferiority *vis-à-vis* Iraq as to wet the appetite of Baghdad to try and unite the two Ba'th republics by force. Jordan, in turn, must be spending its meagre resources on arms not so much against Israel as against Syria and Iraq.

Last but not least, some key Arab adversaries of Israel have had their share of worries with non-Arab neighbours on their flank. Syria had ample troubles with Turkey from the 1940s to the 1960s. Egypt and Sudan were at odds during Sudan's first years of independence. Above all, ever since the Iranian revolution, Baghdad has been at loggerheads with Tehran. As the Iran–Iraq War demonstrated with such devastating effect, such a conflict not merely added fuel to the Arab–Israeli arms race but clearly over-

---

[23] F. Ajami, *The Arab Predicament* (Cambridge: Cambridge University Press, 1981).

shadowed it. This had its impact on Iraq's neighbours, which, in turn, had an impact on the Arab–Israeli arms race and vice versa.

Turning to the future, this dynamics is as pertinent to the issue of non-conventional weapons as it has been in the past. The main reason why Israel has so doggedly avoided going public with the bomb has been its fear that, by doing so, it would be losing rather than gaining in its effort to offset the Arab quantitative edge. With a 'bomb in the basement', Israel has an effective deterrent against a massive breach of its territory by Arab land forces and/or Arab use of crude non-conventional weapons of the chemical kind. Official Arab statements since the early 1970s confirm that the Arabs assume that Israel is a nuclear power, as did Arab conduct in the 1973 Arab–Israeli War, in which the Egyptians did not seriously try to recapture the entire Sinai Peninsula, whereas the Syrians halted their advance on Israel's border even when they could have penetrated deeper.[24] Indeed, even Saddam's decision during the 1991 Gulf War not to use Scuds with chemical warheads, which he had in his possession, against Israel was, according to Israeli observers, an acknowledgement of the viability of Israel's non-conventional deterrent.[25]

Were Israel to go public with the bomb, it would be impossible to deny its adversaries access to the same weapons. If Iraq were not denied access to nuclear weapons, the result could only be one of two equally disastrous outcomes: either an unstable nuclear balance in which he who hits first survives, or a stable nuclear balance (with second-strike capabilities) in which the added strategic value of Israel's 'bomb in the basement' would have been erased. In the former scenario, Israel would either hit first and be damned for ever or wait to be hit by the Iraqis and be annihilated. In the latter scenario, strategic nuclear weapons cease to be usable and the main arena of Arab–Israeli confrontation becomes the conventional one *where the Arabs have an inherent advantage*.

The obvious conclusion of the preceding discussion is that Israel's only long-term hope—and *ipso facto* the only long-term hope for a stable equilibrium in the Arab–Israeli arms race—lies

[24] For a detailed survey of Arab statements and commentaries on this issue which bear this point, see M. Steinberg, 'The Israeli Nuclear Challenge in Arab Eyes', in Klein and Steinberg, *The Arab World in the 1980s*, 63–80.

[25] See, e.g., S. Feldman, 'Israeli Deterrence in the Test of the Gulf War', in Jaffee Center for Strategic Studies, *War in the Gulf*, 170–89.

in a regional system of arms control in which the ability of the Arabs to acquire non-conventional weapons, as well as a mass conventional firepower, is curbed by third-party (especially great power) intervention. Such a system cannot antecede a multilateral settlement of the Arab–Israeli conflict. But it ensures that, if and when such a settlement is worked out, the Israelis will ultimately agree to some form of inspection and control over their own nuclear programme.

# PART FOUR

# ARMS CONTROL

# 10

# Arms Control and Supplier Restraints: A UK Perspective

## WILLIAM HOPKINSON

ARMS control is a process which has undergone very radical change since 1990. In essence it has become, in the East–West context, dialogue, rather than the counting of weaponry—though with the possibility of unilateral action on reductions. In the North–South context the main concern has become non-proliferation.

As regards supplier constraints, there are agreements, such as the NPT and BTWC which limit proliferation whilst being recognizable as arms-control arrangements. Under these, the parties agree not to supply, nor to seek to acquire, the capability concerned. On the other hand, there are also a number of regimes, not based on treaties, which seek to impose constraints upon the supply of weapons of mass destruction, or the potential to create them, without any link to formal arms-control treaties. Classic examples here are the MTCR and the Australia Group. In essence these are arrangements between supplier countries to limit and control transfer to others by national legislation.

Certain new potential restraining regimes, or contributions towards such regimes, are under consideration at the moment—for example, the discussions amongst the five permanent members of the UN Security Council. These may be expected to focus mainly on conventional weaponry, whose proliferation is not otherwise generally covered, rather than on weapons of mass destruction. However, we can expect to hear calls for restraint in all forms of proliferation whenever statesmen are gathered together for summits or multilateral exchanges.

In this chapter, I shall consider arms control and supplier constraints in turn, and attempt to provide a UK perspective on the

This chapter represents the author's own views, and is not a statement of UK government policy.

problem of non-conventional-weapons proliferation in the Middle East.[1]

## ARMS CONTROL

Arms control may be understood as being essentially concerned with measures, undertaken by those on whom they directly bear, to limit, reduce, or forgo certain weapons, or types of weapons. They may be highly structured, with complex definitions and counting rules (e.g. the CFE Treaty or START) or more general (e.g. the BTWC or NPT). In addition, and particularly in the East–West context, arms control has included some very successful measures to increase transparency, build up confidence, and limit deployment of armed forces; these may be referred to as confidence-and-security-building measures.

The early 1990s have seen some remarkable and far-ranging arms-control achievements. The CFE Treaty, though in many respects now overtaken by political developments, was a major contribution in its time to European security, and far surpassed the earlier attempts to limit conventional forces in Europe. A good case can be made that its success owed a great deal to confidence-and-security-building measures, building up from the very modest 1975 Helsinki measures to the 1986 Stockholm Document, and providing a sound basis of trust for more substantive measures. Certainly, that process will be the basis for future developments in European (collective) security. START was a long time in arriving and will produce only relatively modest reductions; nevertheless it was a solid piece of work. There was a successful Review Conference of the BTWC in the autumn of 1991, and a CWC was agreed in 1992.

Against that background, it would be possible to write at some length on UK approaches to arms control in the context of the Middle East. The essence would be that arms control has been linked with, indeed at some stages was a vital component of, the radically improved East–West relations, and the fundamental reordering of European security. It would be possible to point also

---

[1] For the purpose of this Chapter, the Middle East may be considered as extending from Libya to Iran, but excluding Turkey. This may not make good geography, but it comprises an area which makes sense in terms of political and security concerns over proliferation and arms control.

to the 1987 Intermediate-range Nuclear Forces Treaty—a treaty which removed an entire class of what would be called in the Middle Eastern context weapons of mass destruction, and which also introduced what was then an unprecedented degree of intrusive inspection. The involvement of many Middle Eastern states in the BTWC Review Conference could be brought out, and reference made to the chemical-weapons negotiations, now at last, perhaps, entering their final months. Biological and chemical weapons are certainly major Middle Eastern concerns, and, with all the advantages which came to Europe from the other negotiations mentioned above, it would not be unreasonable to say that arms control is good, and the Middle East would manifestly benefit from something of the same process.

Alas, matters are not so simple. There was in Europe, at the crucial stages of negotiations, a bipolar system in which the notion of balance could take hold. No such idea could easily be applied in the Middle East, either to conventional or to other weapons. The area is characterized by multiple and overlapping regional problems. There is Israel–Palestine; there is the rivalry between Syria and Iraq; there is Iran; and there are other Islamic nations, outside the area, but with at least some concern about what is happening there. Moreover, defining the Middle East to include Libya brings in a set of other potential problems with Western nations. Since arms control should, in principle, be undertaken to augment security, it would be necessary to ask whose security would be increased by the process, and how. The answer would not be clear.

That having been said, some things would certainly add to the security of the nations of the area. A total ban on weapons of mass destruction could hardly be said to be anything other than an increment to security—if all could be certain that the ban was effective, and if no nation felt the need to have, say, nuclear weapons as its ultimate guarantee of national survival. Unfortunately the former condition would involve a degree of transparency which is not immediately available, while the latter leads straight into the thorny thicket of Israel's place in the Middle East.

A ban on all ballistic missiles would also be desirable, but that too would run into difficulties. Having disposed of those systems clearly discernible as ballistic missiles, there would need to be reassurance about the future, including about the non-conversion of, say, air-defence missiles and space vehicles to ballistic purposes; again the openness to get to that position is not yet present. However, one important characteristic of arms-control regimes concerned with weapons of mass destruction is that they are often not posited on numerical party, and at the least, one must consider the applicability of the BTWC and CWC to the Middle East.

Moreover, questions of balance apart, there are some lessons from the conventional-arms-control process in Europe which could usefully be applied in the Middle East. At any rate, it is necessary to understand the limitations of the European lessons. The objective of the CFE negotiations was to strengthen stability and security in Europe by eliminating, as a matter of priority, the capability for launching surprise attack and for initiating large-scale offensive action. One could hardly quarrel with that as appropriate also to the Middle East. The objective was to be achieved via a lower level of armaments, and there was rapid agreement that certain types of equipment had to be included to achieve the objectives (tanks, artillery, armoured vehicles).

The CFE Treaty was essentially a bilateral process—giving each side the same could be presented as fair and reasonable. No such symmetry exists in the Middle East. Moreover, the strategic situation of some states is such that almost no level of holdings of (say) tanks by their neighbours would guarantee security. There is little strategic depth in Western Europe, but there is much more than in the smaller Gulf states, or in Israel. In those circumstances, anyone with a few divisions may be able to seize and hold ground. A simple translation of the CFE Treaty is, therefore, not a starter.

On the other hand, there are lessons to be learnt from the confidence-and-security-building measures which may be more useful. The beginnings at Helsinki were modest, the players deeply suspicious of each other, the ability to check the information given very limited, and its military utility not very high. Nevertheless, a process was started which, even leaving aside the benefit of having contributed to the CFE Treaty, produced the block of substantive and useful measures incorporated in the 1990 Vienna Document. If the states of the Middle East could have that document as the basis of their politico-military relations, the world would be a safer place, and a considerable increase in their security would have been obtained.

Unfortunately, the Vienna Document cannot simply be dropped into place in the Middle East. As with almost all arms control, it is almost always the case that, when you need confidence-and-security-building measures, you cannot get them; when you can get them, you do not need them. However, some modest exchanges of information, some exchange visits, some discussions of security concerns could start a worthwhile process. It would certainly be useful to seek to prepare a modest package on those lines for discussion as part of the Middle Eastern peace process.

Arms control is an aspect of security policy, and should not be considered in isolation—hence the need to interweave work on confidence-and-security-building measures in the more general process. Work on biological and chemical weapons is already in train in other places, however, and seeking to weave that other work into the Middle Eastern peace process would probably over-complicate matters. The British, and indeed general, objective is to eliminate all biological and chemical weapons; only if the chemical-weapons negotiations in Geneva fail would there be scope for a specific Middle Eastern regime. Moreover, there is a general reluctance to press for measures which affect one region, rather than being of global application (though it may be necessary to arrange, as part of a Middle Eastern process, that the states of the region move together in applying the general obligations of the various conventions on non-conventional weapons). Most significant Middle Eastern states of concern are also involved in the chemical-weapons negotiations, either as Conference on Disarmament members or as observers, but it would be very difficult, in the context of the chemical-weapons negotiations, to pursue concurrently a different regime for the Middle East—whether that was described as more effective, or discriminatory.

As regards biological weapons, a number of significant Middle Eastern players were involved in the Third Review Conference, even states which are not formally party to the BTWC. It is to be hoped that, both as a result of the Review Conference and as a response to general world opinion on the unacceptability of such weapons, those signatories who have not ratified will now do so, and will faithfully apply the biological-weapons confidence-building measures and more constructive proposals to help increase transparency.

That is the good news—that these general and world-wide

conventions are attracting interest from the Middle East; and, to the extent that they provide useful frameworks, there will be a positive impact there. However, as will be seen below, there will be limits to how far these regimes really meet the stringent requirements of preventing proliferation. Indeed there is an argument that they could even aid it—via member states claiming that, as parties, no supplier-regime controls should apply to them. A similar problem could arise with the NPT. Iraqi actions have given prominence to the need to tackle the issues, but ensuring general adherence to the NPT and faithful application of its provisions is far from simple.

It is not clear what the outcome of the Iraqi adventure will be in terms of safeguards, control mechanisms, and so on. Certainly, events have demonstrated that the current regime is not completely effective; whether any significantly more effective regime could be devised, or at any rate one which was generally acceptable, remains to be seen. And, whilst it is reasonable to discriminate against the present Iraqi regime, which has been caught in flagrant breach of international obligations, that cannot become a general rule; indeed, it cannot continue definitely against the country of Iraq. We must not create another Versailles problem. Iraq and the Iraqi people will bear the burden of the recent unsuccessful war for many years, and that is unavoidable. It would not, however, be appropriate to expect a successor regime, which genuinely sought to meet its obligations, and to be a responsible member of the world community, to bear a greater limitation on its sovereignty and security than its neighbours.

To round off on arms control, it is necessary to consider missiles. Some types could in principle be abolished, by mutual consent, along the lines of the INF Treaty. Specifying the parameters which might apply, and attempting to differentiate between offensive and defensive missiles, would not be easy. The likelihood of all or even most Middle Eastern states being able to agree what categories should go seems remote, given the mixture of ranges and degrees of sophistication, and the uneven spread, particularly of the latter. Even more remote would seem to be the possibility of limiting numbers at levels above zero; how could one find a fair balance?

All in all, therefore, classic arms control has relatively little to offer towards the resolution of the Middle East's non-conventional-weapons proliferation problems. On the conventional side, confidence-and-security-building measures would be desirable, but negotiated limits on major systems are not likely. One aim should be the proper observance of extant treaties; beyond that it is likely that there will need to be supplier constraints, with the acknowledgement that they are unlikely to be palatable to those on whom they bear—even though any one Middle Eastern state would benefit from constraints which would limit would-be proliferators amongst its potential adversaries.

## SUPPLIER CONSTRAINTS

Supplier constraints essentially involve those who are often rich, often Western (in political orientation), and usually secure saying 'no' to others who are poor, Third World, and insecure. Moreover, they involve not merely saying 'no, you may not have weapons of mass destruction', but also 'no you shall not have certain knowledge and technology'. Whilst most states, like all men of goodwill, may agree in principle on the desirability of preventing proliferation, that background means that there will be tensions between developed and developing countries. Nations which see, or allege that they see, a pressing current economic need to develop new technologies which may have military implications will not welcome opposition from those who already have the necessary industrial base.

Moreover, there are, of course, even amongst Western allies, considerable differences over the approaches appropriate to particular nations. For example, the US view on the desirability of collaborating with Israel with high technology suitable for missile applications might well be rather different from that of European nations. France and Italy, because of past imperial connections, have relationships in the Middle East which could lead to their seeing real difficulties in denying technology to (or, as would be necessary in some cases), academic or educational links with) certain countries. Within Europe the development of the European Community may iron out some such problems, but making a multiplicity of supplier constraints work effectively is going to be a demanding task for many years, and that is probably truer of the Middle East than of any other region. COCOM was a successful regime because there was general agreement about its necessity.

The Soviet Union was perceived as a major military threat; appropriate steps were necessary to prevent its being strengthened. That philosophy—the potential adversary—cannot simply be transferred into North–South, or West–Middle Eastern relations.

A comprehensive look at supplier constraints would need to examine the regimes for controlling the exports and technology transfer rules for technologies for biological and chemical weapons, and the MTCR, as well as the nuclear area. All of these are subject to review or expansion at present, not least because of the discoveries of Iraqi developments. (Meanwhile, work is going ahead on what is effectively a supplier regime for conventional weaponry—with the five permanent members of the UN Security Council.) Such arrangements, whereby a number of suppliers (preferably all significant ones) come together to try to agree a common approach, lies at the heart of the current process of attempting to limit proliferation. All responsible states have an interest in presenting the proliferation of nuclear, chemical, and biological weapons. Most would take the same view over missiles—though here attitudes vary because of the potentially defensive applications of some systems.

A full examination of the nuclear regime would be inappropriate here, on the basis of the material currently available. Full lessons have still to be drawn from the Iraqi experience, whilst a new and greatly complicating factor has been the dissolution of the Soviet Union. There has long been considerable praise for the IAEA safeguards; indeed, before the disclosures from the UN Special Commission work, the refrain was sometimes heard that the regime should not be disturbed, because no better one could be put into place. It is now manifestly clear that the safeguards regime is not enough on its own to stop a determined would-be proliferator.

The ability to inspect frequently, at short notice, will clearly be required to make a more effective regime, but the difficult questions are to what facilities, and how often. The ideal answer would seem to be 'anytime, anywhere', but that does not seem likely to be a saleable formula. That is not, of course, to say that (improved) safeguards are necessarily useless; proliferation would be easier in the absence of IAEA controls, and may be so difficult with them that only the most determined will persist. We should certainly not let perfection become the enemy of the achievable good. (One proliferator is certainly better than twenty!) The Nuclear Suppliers' Group has just been revived, and is carrying forward work on dual-use items, and that is certainly to the good. That said, dual use, and the extension of civil plutonium programmes, will continue to be a major potential source of leakage for the future.

As regards biological and chemical weapons, the Australia Group now consists of twenty-two states, essentially Western, at least politically, but including some neutrals. They have agreed on a list of fifty chemicals, in effect potential chemical-weapons precursors, the export of which needs to be closely controlled. They are also expected to move forward in the near future on controlling biological-weapons materials—dual-use potential being the major difficulty there. Regular contacts, exchange of information, and experience are making the Australia Group an effective mechanism. However, continuing technical advances, and the spread of skills and infrastructure to new areas, means that the regime will never be leak-proof. And, in the Middle Eastern context, basic chemical weapons are well within the capability of a number of major states. Supplier constraints will, therefore, not remove the threat of chemical weapons from the region, and the technical difficulties of detecting illicit biological-weapons activity mean that the determined proliferator is unlikely to be stopped, even with much more openness than is at present to be found there.

Nor is it clear that it will be possible to remove the missile threat from the area by supplier controls, given the number of (often rather crude) systems already there. The failure of the Condor system demonstrates that more sophisticated developments may be seriously handicapped, but the ability to drop something crude on a neighbour's city, with a biological or chemical warhead—even if not a very effective one—is already there and will not be removed by non-proliferation rules. Stocks of missiles are already high, and there is certainly the indigenous capability in many states to make crude missiles (and, in Israel, the ability to make sophisticated ones).

The MTCR with its payload–range criteria of 500 kg–300 km—originally framed so as to prevent the development of a nuclear–capable missile—could never have been the total answer to missile proliferation; it is now for consideration whether the criteria should be changed. Of equal importance, however, is whether wider adherence to it can be achieved. The Soviet Union's expressed willingness to abide by the rules (stated in 1990) was a most welcome step forward; however China and North Korea have still not given any bankable undertaking, and both are already suppliers to the Middle East. Assuming that all these potential suppliers are prepared to stand solidly with Western MTCR members, then the spread of long-range, more accurate missiles could be limited for a time. However, since Israel already has, it would appear, both accurate guidance technology and longer-range vehicles, others will continue to seek and over time acquire them too, unless there is a genuine settlement.

### REGIONAL GLOBAL MEASURES

UK objectives on non-proliferation in the Middle East are, of course, part of wider objectives of general application. In essence, the arguments as to whether we should work exclusively on the Middle East or whether we must build on world-wide regimes are fairly simple. Middle Eastern nations will not tolerate, at any rate in the longer term, what they would see as a discriminatory regime. It is, in any case, doubtful whether the United States could contemplate an effective regime which operated against Israel, and, to the extent that it did not, the reaction from Arab states would be all the stronger. On the other hand, there is a real security problem which needs tackling with great urgency. The convoy should not have to wait for the slowest ship because of some concern about particular aspects of non-proliferation regimes from states outside the region.

If there were a genuine general consensus in the region, one would certainly not wish to block progress because of concerns, however legitimate, over (say) Korea or the India subcontinent. In the realm of classical arms control, the 1990 proposals by Egypt for the elimination of all weapons of mass destruction, with verification, would be a desirable objective. If the conditions for that were fulfilled, well and good; but if not, the question would be whether suppliers were willing to take a more stringent attitude with Middle Eastern countries than with others. They might be, but there would be grave risks. There is little point in a stringent regime for the Middle East, amongst existing suppliers, if there is the possibility of evasion by transfers from non-members of the suppliers' club outside the area. Thus, unlike confidence-building, or even straight arms-control measures, which could certainly be, under some circumstances, effectively based on a regional settlement, a supplier regime, at any rate for weapons of mass destruction, probably needs to be world-wide.

That is not to say that the particular urgency of tackling proliferation in the Middle East is unrecognized. Indeed, were the case not such a difficult one, there might be virtue in introducing measures there first, leaving rules of wider application to be developed later. What has received less attention so far than supplier regimes is steps to reduce the demand side of proliferation, to damp down the desire of states to acquire new weaponry. That seems likely to be a vital aspect of the necessary political process in the Middle East; it is something in which confidence-and-security-building measures, and transparency in particular, should have a role.

That, however, points up the difficulty—the classic difficulty—of arms-control arrangements. If we are speaking of measures accepted by those to whom they are applied, then there will have to be a considerable advance in security before that can happen. For a Middle Eastern state to be willing to forgo the option of developing a weapons system, it must be tolerably clear that it is not at risk from its neighbours' stealing a march; and it will have to be very clear that any concessions which it makes on openness and transparency will not unilaterally disadvantage it. The problems of generating mutual confidence between (say) Syria and Israel are manifest, which is why specific Middle Eastern arms restraints are more likely as a result of, or as part of, a general Middle Eastern settlement, rather than as the first steps to a world-wide regime.

### TECHNOLOGY-TRANSFER REGIMES

More generally, a major potential source of difficulty between the supplier groups and the less-developed nations is likely to arise from the thesis that with, the establishment of an effective CWC, and perhaps an improved BTWC, all that is necessary to have ready access to technology, equipment, and substances is to be a paid-up member of the convention, ratified, and open to whatever inspection or confidence-building regime there may be. This emphatically seems unlikely to be good enough; within the foreseeable future neither the CWC nor the BTWC will have the sort of intrusive verification arrangements which could give reasonable assurance about certain of the regimes which are likely to be found in the Middle East.

Suppliers will have to continue to collaborate, and there will, in effect, have to be lists of substances, apparatus, and countries subject to special controls. The preparation of these will, in some measure, depend upon intelligence, for there will be a need to consider the export not only of lethal substances (on the whole rather unlikely), but also of apparatus for producing substances, of skills and knowledge about design and production, and indeed even of personal skills. Those operating the control mechanism will need to consider the likelihood of a particular regime, or sometimes even an individual, abusing any transfers. It will be important to know, when an application for a potentially dual-purpose piece of apparatus is received, who is really seeking it; what he is really up to; and where its final destination really is. Even in these prodigiously changed days, intelligence-sharing, even with long-standing allies, is a difficult business. The United States has unrivalled facilities, but it will probably be reluctant to share with other Western nations, let alone (say) China, all that it knows. There is a limit to what can be expected, particularly in so sensitive an area as the Middle East.

It goes against the grain to say that there will have to be restrictions on, for example, academic interchange. However, there are certainly regimes in the Middle East which are prepared to have scholars working in the West on technologies and processes which would have implications for the production of weapons of mass destruction. In other words, questions will have to be asked about Doctor X who has applied to come to study at Oxford; about the institution which he represents; and about the long-term potential of the area in which he wishes to work. At a time when there is increasing economic importance attaching to areas like bio-technology, that will give rise to considerable strains.

Similar problems will arise over entry into commercial undertakings. It now seems clear that Iraq had a widespread policy of acquiring access to technology by buying companies in the West.

At a time of ever-opening markets, and of former Warsaw Pact and Soviet states seeking hard currency and business partnerships, there will be a major problem in ensuring that Middle Eastern proliferators with hard currency available do not, either directly or through proxies, enter into arrangements which give access to technologies with military applications. All in all, this area will give rise to continuing difficulties, especially as the NPT and the BTWC encouraged peaceful co-operation as a quid pro quo for restraint on military application of technology. Unfortunately, it now appears that no simple division can be made; many, if not all, technologies have a dual use; tensions between suppliers and potential acquirers, whether malevolent in intention or not, will be a continuing feature.

A deepening of the MTCR, NPT, Australia Group, and similar regimes may create a dense web which raises the cost of proliferation to the would-be proliferator. As noted above, however, that will not be effective if he can simply go to an alternative source of supply. As technology spreads, there will, inevitably, be more countries capable of assisting proliferation. Supplier regimes must, therefore, expand, so far as they reasonably can, to catch significant potential suppliers—or at least those who are of the right inclination as regards preventing proliferation. (To include those whose hearts are not committed would be worse than useless; it could prove positively harmful, by alerting regimes interested in engaging in or facilitating proliferation as to what is afoot, and how best to evade controls.)

There is a real risk of paralysing any regime by making it top heavy: too many rules, and too many members, will mean only that the machinery will break down. Nevertheless, progress is possible. The steady increase in the controls applied by the Australia Group illustrates the deepening of a regime; the expansion of the MTCR—perhaps to take in the Soviet Union's successor states—is an example of a broadened membership. In all cases, the regimes will need to be both widened and deepened: widened in so far as more and more nations acquire the skills and technology to make them potential exporters; deepened as more and more technologies are evolved which could have military implications.

Moreover, there will come an increasing interface with other regimes, not designed to deal with weapons of mass destruction.

For example, the P5 are seeking to limit (*inter alia*) conventional proliferation, and weapon build-ups. As part of that they will need to consider missiles with particular care—for these can convey significant political and military advantages. They are certainly part of any up-to-date conventional armoury; they also have strong links with non-conventional systems. The impact of the P5 discussions on the Middle East may, in the initial years at any rate, occur more through helping to create a consensus about the nature of the problems and concerns than by agreeing a strict policy on limiting arms transfers; nevertheless, this forum will probably come to be a major aspect of supplier action for the region.

### THE UK PERSPECTIVE

The UK government's perspective on these issues will be influenced by the fact that the country is involved in several security fora, fora which are, moreover, developing and changing rapidly. A founder member of the Western European Union and NATO; committed to the European Community, although not always in the majority on how its security dimension should be developed; having a special relationship with the United States, at any rate as regards nuclear and intelligence matters, the United Kingdom is also a significant player in other relevant fora. For example it is an active member of the Conference on Disarmament in Geneva where the CWC was negotiated; a co-depositary of the NPT and BTWC; a country which plays a full role in various non-proliferation fora such as the Australia Group and the MTCR. Given the need for the United Kingdom to evolve a coherent foreign and security policy, taking account of national interests and yet enabling progress to be made in all these different fora, there is unlikely to be room for a simple UK position or perspective, on anything beyond the general objectives of enhancing stability, eliminating proliferation, and, in short, encouraging virtue.

Fortunately, perhaps more than most countries, we remember Lord Palmerston's declaration: 'We have no eternal allies, and we have no perpetual enemies. Our interests are eternal and perpetual, and those interests it is our duty to follow.' It is not quite certain that nowadays the United Kingdom has perpetual interests. Be that as it may, it has consistently argued for a world-wide, verifiable CWC; for seeking general improvements to the BTWC; and for maintaining the NPT. These make for a cohesive, coherent package, and one which fits, moreover, with other aspects of the UK security policy.

In that context, it may be appropriate to deal with one point concerning the NPT, regarding the United Kingdom's own possession of nuclear weapons. The United Kingdom has consistently grounded its policy on the fact of the stabilizing effect of nuclear weapons in the East–West context; it has not argued that they are generally stabilizing. It has been no part of the UK position that there is a general case, unrelated to context, for the possession of nuclear weapons. Moreover the United Kingdom, being mindful of the obligations of Article VI of the NPT, has never ruled out participation in negotiations at an appropriate time. The point is simply that most of the ground to be covered for Article VI is clearly for the United States and the Soviet Union's successor states; given that fact, and the contribution that nuclear weapons have made to stability in certain very specific circumstances, there is no inconsistency in the United Kingdom seeking to limit proliferation whilst remaining a nuclear-weapon state.

A more general UK concern and perspective on arms control and proliferation is with the need for deeply intrusive observation of activities, if breach of obligations is to be established. Deeply intrusive inspections can uncover the development of some weapons of mass destruction. Both nuclear and chemical weapons require a considerable amount of infrastructure and input. (Biological weapons requiring significantly less of both may be different.) Deployable missiles can also be found by such intrusiveness. Without deeply intrusive measures we, and more importantly the nations of the region, will not know whether proliferation is going on. We should, in the context of trying to get a general Middle Eastern settlement, undoubtedly press for as much openness, as much exchange of information, visits, etc., as possible. But advice from the West, on such matters will not generally be welcomed in the area; and again there will be the problem of Israel; will the United States be able to put appropriate pressure on that state to co-operate with its neighbours in measures of openness?

## CONCLUSION

To sum up, there seems little scope for classic arms control, of the CFE or START type, in the Middle East. A dose of confidence-and-security-building measures, which might be the precursor of weightier developments, would be welcome—but that must be as part of a general settlement. Transparency and openness are of the essence.

Existing supplier-constraint regimes have particular problems. The MTCR was originally designed to cover missiles capable of delivering a simple nuclear weapon—the criterion was a 500 kg payload over three hundred kilometres. In terms of constraining weapons of mass destruction that is clearly not sufficient. Not only is there a trade-off (albeit not a simple one) between payload and range, but a 500-kg payload would provide an enormous capacity for biological-weapons delivery, as well as a not insignificant one for chemical weapons. (Indeed, almost any biological-weapons payload could, with appropriate fusing technology, make any missile into a weapon of mass destruction.)

Missiles are especially difficult, in as much as technologies or components or indeed whole missiles designed for one purpose can be adapted to another. Many of the technologies for space-research vehicles can, for example, be readily applied to ballistic missiles. And even defensive missiles, designed for anti-aircraft or anti-missile roles, can, as has been seen in both the Middle East and Korea, be redesigned to have a ballistic role, even if not the most efficient one. Moreover, as the 1991 Gulf War showed, for a weapon to be a terror weapon, it need not necessarily be very long-ranged or accurate. Given the large availability of Scud-type weapons throughout the Middle East, the possibility of getting the genie back into the bottle and preventing proliferation of simple weapons of terror through the MTCR seems very limited. What may be possible is to prevent the introduction of weapons of higher degrees of sophistication, longer range, etc., if the suppliers, either formally within the MTCR, or by agreeing to go along with it, can be brought into line. The MTCR, if it is followed by such states as China and North Korea (even though not members), could slow down the spread of the longer-range missile threat, but missiles in the Middle East are probably something with which the world is going to have to live.

On chemical and biological weapons the story is grimmer. There is widespread knowledge of chemical-weapons processes in the Middle East, and many states are either chemical-weapons capable, or able to become so, at any rate in a primitive way, very quickly. Biological weapons may not be so widespread, but again the knowledge is abroad, and states other than Iraq may well have programmes. That said, the Australia Group has done useful work in tightening up the chemical-weapons lists and in adopting new biological-weapons ones. Again these will be useful steps, but not sufficient. Many legitimate chemicals can be fairly readily adapted or adopted for military use; the number of countries able to supply them is likely to increase. Again it may be possible to prevent more sophisticated weaponry or potential weaponry spreading, but there will remain a major problem with crude weapons, which could be particularly effective in wars of cities in the Middle East.

Israel is said to have a nuclear capability; Iraq was fast approaching it. Even if its infrastructure is now destroyed, the individual skills and knowledge will be able to migrate elsewhere in the Arab world. Without intrusive measures of some sort, confidence that further nuclear proliferation is not in hand will not be forthcoming. In the light of the experience with Iraq, clearly the nuclear regimes will have to be tightened up; IAEA safeguards as hitherto practised are not a sufficient restraint; what *would* be sufficient in the case of a determined proliferator is not immediately clear, although various ways of tightening up the systems are under consideration.

So what is the perspective of this chapter on arms control and supplier constraints in the Middle East? First, arms control is a good thing, and some arms-control regimes will certainly have a role to play there—for example, on biological and chemical weapons; however, a 'European'-style breakthrough is not likely. Secondly, supplier constraints are most certainly needed; wider and deeper regimes are required, but as such they are not going to make the Middle East a safe place. On many of the systems which are a cause for concern, the genie is already out. Possibly the way forward, and the only source of confidence that the problem will be properly addressed, is with the development of inspection regimes, internally developed in the Middle East. A regime sufficiently intrusive to stop biological weapons would be very difficult to devise; only one which is much more intrusive than anything that is likely to be acceptable to (say) Israel and Syria could stop nuclear proliferation, but one must look for a start, and an increase in confidence could lead to a reduction in the demand for proliferation.

In short, the solutions to the region's problems lie elsewhere than in arms control or supplier constraints. Let the regimes be strengthened, but as contributions to a wider settlement.

# 11

## Arms Control and the Arab–Israeli Peace Process

### GEOFFREY KEMP

THE conference on Middle East peace held in Madrid at the beginning of November 1991 and the initiation of bilateral negotiations between Israel and its key Arab neighbours have highlighted the agenda of issues that need to be resolved before a comprehensive Arab–Israeli peace will be possible.[1] The most difficult substantive questions relate to geographic borders, political legitimacy, the right of return of the Palestinians, Israeli settlement activity, reparations, and compensation, access to resources such as water and the status of Jerusalem, as well as relations between Israel, a Palestinian entity, and the Arab countries, and security questions, including arms control and mutual-force and weapons limitations. The focus of this chapter is on the relationship between arms control and the peace process.

Arms control covers a wide variety of initiatives—unilateral, bilateral, and multilateral—and can cover the gamut from informal, confidence-building measures such as red lines establishing military ground rules between adversaries, to formal, multilateral treaties to eliminate entire classes of armaments, such as nuclear and chemical weapons. The peace process refers to a complicated series of negotiations between adversaries that passes through at least three distinct, but integrally linked stages. These can be identified as: pre-negotiations, negotiations, and post-negotiations. At each phase different agreements are necessary to

---

[1] This chapter draws on material first published in G. Kemp, *The Control of the Middle East Arms Race* (Washington DC: Carnegie Endowment for International Peace, 1991), Ch. 8. Additional material has been added to take into account the Madrid peace process and new information about nuclear proliferation in Israel, Iraq, and Iran.

move the process forward and different arms-control initiatives are appropriate.

It is anticipated that arms-control issues will eventually be on the agenda of both the bilateral and multilateral Arab–Israeli talks, though it is unlikely that they will be priority items precisely because of the sensitivities involved. Therefore, it is worth examining arms control as a confidence-building measure that might be implemented prior to the issues being formally addressed in the negotiations.

## An Improved Climate for Limited Arms Control?

The Egyptian–Israeli Peace Treaty, finally consummated in April 1982, had its origin in agreements negotiated between the United States, Israel, and Egypt between 1974 and 1979, and in Egyptian President Anwar Sadat's historic trip to Jerusalem in 1977. One reason the treaty has survived is because there exist complex arms-control agreements limiting military activity in the Sinai Peninsula and along the Egyptian–Israeli border. If Egypt and Israel were the only parties to the Arab–Israeli conflict, by now they might have begun discussions on regimes to restrict weapons of mass destruction and their delivery systems, including advanced conventional arms. However, due to Egypt's wider involvement in the Arab world, and the absence of substantive peace talks, it is premature to expect substantial progress on such issues at this time. This is not to diminish the role of international efforts to negotiate a chemical-weapons or nuclear-weapons ban. But, in the last resort, the regional parties are unlikely to accept global weapons restrictions until peace has been achieved.

This creates a paradoxical situation. Arms-control initiatives between Israel and the Arabs to ban or limit major items prior to, or even during, an ongoing peace process are unlikely to succeed. However, the political problems of reaching a peace settlement and deciding where Israel's security ends and Arab insecurities begin are so complex that a decision to postpone major arms-control initiatives encourages a continued arms race. The practical approach to reconciling these two positions must be to pursue confidence-building measures prior to Arab—Israeli reconciliation, and to accept that substantive progress on resolving the difficult issues of nuclear and chemical weapons and long-range delivery systems must wait until the political environment has improved.

The overall Arab—Israeli conflict is still in the pre-negotiations phase; that is to say, while all parties have shown some interest in a peace settlement and some have attended the Madrid meetings, there is no agreement as to the next steps and how to include regional countries that were not present at Madrid. In this context, limited confidence-building measures that already exist might be expanded and new ones might be initiated to contribute to an atmosphere conducive to future negotiations.

There have been indications, discernible before the 1991 Gulf War, that both Egypt and Israel regard arms-control proposals as a legitimate issue for discussion. There are several reasons for this new attitude. First, given the nature of the arms race, there is concern that, if military-procurement trends continue, the dangers of war and the costs of deterrence will grow. However, while a dialogue on specific security issues under the rubric of arms control might have provided a vehicle for more substantive talks on other sources of conflict—in the best of circumstances, arms-control talks themselves might have been the precursor for wider-ranging peace talks—the Madrid meetings suggest that mutual arms-control discussions must now follow negotiated agreements on more fundamental issues.

Secondly, the dramatic developments in the East—West dialogue on conflict resolution and force reduction have put pressure on regional leaders to take similar steps. Otherwise, in their new spirit of co-operation to resolve regional conflicts, the external powers may initiate regional arms-control regimes that the regional powers find intrusive. While the joint declaration of the five permanent members of the UN Security Council on arms transfers and non-proliferation in London on 18 October 1991 is a bland document with no enforcement mechanism, it is indicative of greater interest in supplier arms control and may portend more effective constraints on arms sales.

Thirdly, in terms of regional politics, it is sensible to be publicly supportive of arms control, which is difficult to oppose in the abstract. Furthermore, advocating arms control can potentially serve each country's interests. For instance, by supporting a ban on all weapons of mass destruction, the Arabs achieve two goals:

they force Israel to be more open about its nuclear-weapons programme, and they help diffuse international criticism of their
chemical-weapons programmes. On the other hand, by agreeing to
participate in regional discussions on weapons of mass destruction, and by supporting a separate conference on conventional
weapons, Israel puts pressure on the Arabs either to enter direct
negotiations or to be seen as the obstacle to progress.

Fourthly, while few political leaders in Egypt, Jordan, or Israel
believe that a full-scale premeditated Arab–Israeli war is imminent,
all agree that the danger of a major crisis, or war by miscalculation, is cause for concern. Thus, while the wisdom of arms
restraints may be questioned, there is a consensus on the need for
measures to prevent wars by miscalculation and to limit the consequences if such wars occur. Whether the Syrian leadership also
holds to these views is not yet clear.

With these new attitudes in mind, there are a number of
confidence-building measures that warrant careful examination in
the hope that a substantive Arab–Israeli dialogue may eventually
be convened. Some of these ideas attracted interest in the region
before the 1991 Gulf War, and might be resurrected at the
appropriate time. They have merit, irrespective of the station of
peace negotiations.

## Measures to Reduce the Risks of War by Miscalculation

Israel, and those Arab neighbours with whom it is still technically
in a state of war, must find ways to avoid miscalculations about
one another's military intentions. There have been two incidents
over the past few years that could have led to inadvertent war: the
first between Israel and Syria, and the second between Israel and
Iraq.

In April 1986 Nizar Hindawi, a Syrian terrorist originally
from Jordan, attempted to smuggle explosives on to an El-Al
flight leaving London's Heathrow airport bound for Israel.
The explosives were detected and removed, and Hindawi, after
approaching the Syrian embassy in London for help, turned himself in to the British police.[2] It remains unclear whether the

[2] H. Cobban, *The Superpowers and the Syrian–Israeli Conflict, The
Washington Papers* (Washington DC: Center for Strategic and International
Studies, 1991), 74.

Hindawi operation was typical of Syrian terrorist operations and was personally supervised by Syrian President Hafez al-Asad, or whether this and other Syrian terrorist acts are an outgrowth of rivalries or misunderstandings within the Syrian security apparatus. Nevertheless, the conclusion must be that, if the El-Al plane had been destroyed with over two hundred killed, Israel would almost certainly have taken harsh military measures against Syria that could have led to all-out war. If it is assumed that Asad did not want a war and was unaware of the Hindawi plot, then one way to avoid future incidents would be some form of tacit Israeli—Syrian co-operation to strengthen intelligence-gathering on terrorist operations that could be damaging to both countries.

The second event concerned an explosion that took place in Iraq in September 1989, when it was reported that a missile-production facility outside Baghdad had blown up.[3] On 15 March 1990 an Iranian-born journalist working for the London *Observer* who had been caught illegally investigating the site of the explosion was hanged in Baghdad. At that time, it was feared that another accident in an Iraqi munitions plant might be attributed by Saddam Hussein to Israel. Under certain circumstances, this could have caused him to launch an attack against Israel in the expectation that, if he did not, he would lose all his forces in an Israeli attack. Mechanisms to convey information about such a crisis to each side might dispel fears of a surprise or pre-emptive attack. Before the Gulf crisis there was some talk of exploring ways to establish an Israeli—Iraq hot-line, possibly using the good offices of Egypt.

### Transparency Issues

Transparency, a term that has come into vogue in the European arms-control context, refers to that aspect of the CFE Treaty which seeks to remove doubts and fears and build trust by generating an expanding process of openness. By calling for both sides to reveal information on military practices and exercises, the CFE Treaty hopes to develop East—West ties which help remove the misunderstandings and misperceptions that fuelled the Cold

[3] H. Morris, 'Explosion that Led to Ill-Fated Mission', *Independent*, 16 Mar. 1990.

War. A number of proposals have been offered to create this openness, all of which can be referred to as confidence-and-security-building measures. Among others, they include annual exchanges of military information, notification of military activities, and exchanges of annual military calendars.[4]

While superficially appealing, these options have a drawback in the Arab–Israeli context. As long as the states in the region believe that the use of force is legitimate and, under the right circumstances, beneficial, transparency measures can be used for deceptive purposes. The 1973 Arab–Israeli War began with what was announced as a training exercise. On the other hand, during the summer of 1989, Israel was alarmed by joint Iraqi–Jordanian air-training flights along the Jordan Valley. Prior notification might have downgraded that concern.

The most difficult transparency problem concerns Israel's nuclear-weapons programme. The post-Gulf War arms-control debate has focused attention on Israel's nuclear arsenal and regional nuclear proliferation. President Bush's May 1991 initiative offered a cautious but sensible approach to dealing with these issues. By calling for a freeze on further nuclear proliferation in the region rather than outright abolition, the initiative attempted to deal with the threat that the spread of nuclear weapons poses, while also respecting Israel's security requirements. It is plausible that Israel would agree to a halt on further production of nuclear weapons provided it were convinced that the Arab states would not acquire the same weapons.[5]

While greater transparency in Israel's nuclear-weapons pro-

---

[4] Other measures include: risk reduction through reports on unusual and unscheduled activities; contact developments, including visits to military facilities and the establishment of direct relations between military personnel; observation of military activities; prior notification of certain military activities; compliance and verification measures through inspections and evaluations; established communication between capitals; annual implementation assessment meetings; refraining from threats of using force while retaining the right to self-defence. See 'The Vienna Document 1990 of the Vienna Negotiations on Confidence- and Security-Building Measures', 17 Nov. 1990; I. Daalder, *The CFE Treaty: An Overview and Assessment* (Washington DC: Johns Hopkins Foreign Policy Institute, 1991).

[5] See A. Cohen and M. Miller, 'Defusing the Nuclear Middle East', *New York Times*, 30 May 1991; 'Establishment of a Nuclear-Weapon-Free Zone in the Region of the Middle East: Study on Effective and Verifiable Measures which would Facilitate the Establishment of a Nuclear-Weapon-Free Zone in the Middle East', UN General Assembly, Forty-Fifth Session, Agenda Item 49, 10 Oct. 1990.

gramme is demanded by the international community, there are strong countervailing arguments to keep Israel's bomb 'in the basement' and not to disclose too much information about its capabilities. For instance, if Israel were to announce an explicit doctrine of nuclear deterrence and provide details of its nuclear forces—analogous to what is known about UK and French nuclear forces—the Arabs would be under great popular pressure to acquire their own nuclear device. Similarly, it is difficult to see how the United States could avoid some painful choices in its relationship with Israel in view of strict US laws and administration policies regarding proliferation in other countries in the Third World. The publication of Seymour Hersh's book on Israel's nuclear programmes provided a wealth of evidence to confirm the existence of a large, sophisticated nuclear force.[6] Thus it has become increasingly difficult for Israeli officials to deny the existence of these weapons, or for Israeli commentators to refuse to talk about the impact of these weapons on arms-control discussions with the Arabs.

## Red Lines

Red lines usually refer to informal agreements between adversaries concerning limits on the deployment and use of armed forces in given geographic regions. Red lines have been an integral part of both formal and informal force-separation agreements in the Arab–Israeli conflict. They have been used in two ways. First, in the literal sense, red lines can refer to a geographic line or boundary. This line can be agreed upon by the parties and established as a marker beyond which one or both sides will not deploy forces. Secondly, red lines, in a conceptual sense, refer to conditions that prohibit certain actions. These, like the geographic lines, can be mutually agreed upon, or simply laid out by one party.

In April 1976 Syria and Israel reached a secret agreement through US mediation to minimize the chances for confrontation in Lebanon. Syria agreed to Israeli red-line conditions, allowing Syrian military intervention in Lebanon provided that it was restricted to ground forces and that these did not move south of a

[6] S. M. Hersh, *The Samson Option* (New York: Random House, 1991).

line between the Zaharani estuary on the Mediterranean and the village of Mashki in the Bekaa Valley. As part of this agreement, Syria was to respect Israel's legitimate security concerns in southern Lebanon and to avoid air attacks against Christian targets.[7]

Israel and Jordan also have an informal red-line agreement. Since historic attack routes suggest an invading Arab army would attack from the east through Jordan, Israeli military strategists have consistently viewed Jordan's eastern and northern borders as red lines. If Iraqi or Syrian forces entered Jordan in significant numbers, Israeli forces would automatically respond. Understandings between Jordan and Israel on this condition have benefited both countries. For Israel, such a policy bolsters against its deterrence posture. For Jordan, this form of agreement bolsters national sovereignty and protects against Syrian and Iraqi intervention. Israel and Jordan also have informal understandings concerning their shared border along the Jordan Valley. These understandings have dealt mostly with attempts to combat the infiltration of terrorists into Israel, thereby reducing the potential for political tension between Amman and Jerusalem.

An additional example of successful deployment limitations sprang from Israeli concerns about Saudi air-force deployments. Strong Israeli opposition to the sale of US F-15s to Saudi Arabia in 1978 led to a US–Saudi understanding on the deployment of the planes in Saudi Arabia. Sixty-two aircraft were eventually sold, but the aircraft were not to be stationed near the Israeli border. Israel was primarily concerned about the construction of an air-base at Tabuk in north-west Saudi Arabia. Under pressure from Israel, the United States insisted that none of the new planes could be stationed at Tabuk.[8]

## Weapons-Testing Limitations

Similarly, agreements to limit or ban testing of certain weapon systems might also help to build confidence. For example, an agreement by Israel not to conduct flight tests of ballistic missiles

[7] See M. Ma'oz, *Asad: The Sphinx of Damascus* (New York: Weidenfeld and Nicolson, 1988); A. Yaniv, *Dilemmas of Security: Politics, Strategy, and the Israeli Experience in Lebanon* (New York: Oxford University Press, 1987), 60–1.

[8] W. B. Quandt, *Saudi Arabia in the 1980s: Foreign Policy, Security, and Oil* (Washington DC: Brookings Institution, 1981), 61.

would be a significant gesture to its adversaries. If this was paralleled by an Arab agreement not to test ballistic missiles, it would be a useful, reciprocal confidence-building measure. Combined with external restraints on the supply of missiles to the region, such a measure could significantly reduce the potential for offensive missile attacks.

### Multilateral Talks on Chemical, Biological, and Nuclear Weapons

On 16 April 1990 Egyptian President Hosni Mubarak offered an initiative to declare the Middle East a region free of weapons of mass destruction. His proposal, transmitted to UN Secretary-General Javier Perez de Cuéllar, states that: 'It is Egypt's considered opinion that chemical weapons should be dealt with in a comprehensive and global context involving all types of weapons of mass destruction, whether nuclear, chemical, or biological, in order to ensure international and regional security.'[9] Mubarak's initiative called for the prohibition of all weapons of mass destruction from the Middle East, commitments from all regional states to abide by the agreement, and verification measures 'to ascertain full compliance by all States of the region with the full scope of the prohibitions without exception'.[10] In explaining such a proposal, Egyptian Foreign Minister Ismat Abd al-Mujid argued that the Arabs cannot be expected to be 'sitting ducks' and allow the Israelis to have nuclear weapons while the Arabs catch the flak for acquiring chemical weapons.[11]

Israeli officials have expressed a similar desire to rid the region of threats from non-conventional weapons. Prime Minister Yitzhak Shamir stated, '[We] have been proposing to enter bilateral and direct negotiations with any Arab country on the demilitarization of this region of lethal weapons.'[12] The Israelis assert, however, that this must be done in a way that maintains Israel's security

---

[9] See letter 130/90 from Ambassador Amre Moussa, Permanent Representative, Permanent Mission of the Arab Republic of Egypt to the United Nations, to UN Secretary-General Javier Perez de Cuéllar, 16 Apr. 1990.

[10] Ibid.

[11] Quoted in M. Rodenbeck, 'Egypt: Reasserting its Position', *Middle East International*, 27 Apr. 1990, p. 9.

[12] See 'Shamir Comments on Talks', *Jerusalem Television Service*, 16.30 GMT, 13 Apr. 1990, translated in *FBIS-NES*, 16 Apr. 1990, pp. 20–1.

and that assures direct negotiations with Arab countries and direct inspection of one another's facilities.

Other proposals include the suggestion that Israel should unilaterally open up its Dimona nuclear reactor to inspection in return for further Arab commitments not to acquire nuclear weapons. While both of these suggestions would fall short of a comprehensive agreement to establish a nuclear-free zone, they would be regarded as confidence-building measures, especially in the wake of concern about Iraq's flagrant violation of its NPT commitments and increasingly disturbing reports that China has assisted Iran with items that could be part of a bomb programme.[13]

### THE NEGOTIATIONS PHASE

#### Territorial Compromise

Return of territory is seen as the key to an Arab–Israeli peace, particularly parts of the territory known as historic Palestine, or Eretz Israel. This will also be the most difficult issue to resolve during peace negotiations. Territorial compromise will not be possible unless there are iron-clad security guarantees among the parties, analogous to those already in place between Egypt and Israel. These security guarantees must include further force separations, demilitarization, peace-keeping forces, force-deployment limitations, and inspection and verification regimes. There must also be agreements between Israel and the Palestinians concerning the internal security of a new Palestinian regime and its relationship to Israeli security. Since territorial compromise is the key to a lasting Arab–Israeli peace, and since security guarantees are necessary before there can be any Israeli withdrawal from occupied territory, the linkage between these two subjects is of ultimate importance.

#### Territory and Security

Israel's borders with its neighbours following a peace settlement will be influenced by three basic considerations: military relations

---

[13] See A. Cohen and M. Miller, 'Nuclear Shadows in the Middle East: Prospects for Arms Control in the Wake of the Gulf Crisis', MIT Center for International Studies, Dec. 1990. See also R. Jeffrey Smith, 'Officials Say Iran is Seeking Nuclear Weapons Capability', *Washington Post*, 30 Oct. 1991.

with the Arab world, the external and practical day-to-day relations with the Palestinian regime, and the internal-security problem. The issue of external security raises questions about strategic relations between Israel and Syria, and between Israel and Jordan. In the former case, the key issues relate to the control of the Golan Heights and the Israeli–Lebanese border. In the latter case, the key question concerns Jordan's eastern border with Iraq and Israel's likely insistence that Jordan agree to limit Iraqi or other Arab military access to this border (a similar agreement with respect to Jordan's borders with Syria and Saudi Arabia can also be anticipated).

Israeli experience in the 1991 Gulf War suggests some additional lessons concerning the importance of territory. On the one hand, the need for early warning of a missile attack and the forward deployment of missile defences, especially early-warning systems, strengthens the case for not relinquishing the high ground on the West Bank, in the Golan, or forward positions in the Jordan Valley. At the same time, however, the missile threat also suggests that adding a few kilometres to the West–East depth of Israel's border is not going to make much difference with the next generation of surface-to-surface missiles, which are much more sophisticated than the Scud.

Because it sided with Iraq during the 1991 Gulf War, the Palestinian population alienated even the dovish faction in Israel that had called in the past for negotiations with the PLO and for the creation of a Palestinian state. As a result, most Israelis at this time do not believe that Palestinians and Jews can live together. This strengthens the case for more explicit physical separation. Unless Israel is willing to become even more rigid in its control over the territories, or to encourage by peaceful, or violent means the mass emigration of Palestinians, the case for Israeli–Palestinian disengagement under the format of autonomy, self-rule, or even independence (with or without confederation with Jordan) becomes more persuasive.

Israeli hardliners will continue to argue that the creation of a Palestinian state, based on the withdrawal of Israel from the West Bank and Gaza, would inevitably lead to another Arab–Israeli war.[14] They cite the proximity of the West Bank and Gaza to

---

[14] M. Widlanski, 'Current Debate: How Dangerous would a Palestinian State Be? Very Dangerous', *Tikkun* (July–Aug. 1990), 62.

Israel's cities and industries, and the West Bank's strategic advantage of high ground as two reasons to retain the territories.[15] The return of the West Bank would also eliminate Israeli control of the Jordan River as an obstacle to forces attacking from the east.[16]

Such dire scenarios, while technically feasible, must be dismissed as highly unlikely. They assume that a Palestinian state will be permitted to have heavy armaments, and that its borders will be analogous to those of the pre-1967 territories. While this may be Arab wishful thinking, it is not only Israeli hardliners who would reject such an outcome. The vast majority of Israelis— hawk and dove—would never, under any circumstances, agree to a heavily armed Palestinian state occupying the pre-1967 region. To dwell on such scenarios is to miss the point of the debate in Israel: what are the dangers of returning *some* of the territory to a Palestinian regime with strict limitations on its military or paramilitary capacity, together with an intrusive Israeli, UN, or international presence on the high ground of the West Bank and along the Jordan Valley? In sum, the real debate is between those supporting a plan for partial Israeli withdrawal and annexation of certain border areas, and those who believe in no exchange of territory at all.

Concerning the external-security threat, many Israelis believe that the Arab states will eventually upgrade their military potential to rival Israel's both quantitatively and qualitatively, thus leaving Israel with only its territorial advantages.[17] This argument has been temporarily undercut by the defeat of Iraq in the 1991 Gulf War. However, over time, Iraq could re-emerge as a threat. In the meantime, Saudi Arabia's and Egypt's inventories are likely to be upgraded with sophisticated US weapons and, over the horizon, Iran and Pakistan loom as potential military adversaries. Summing up this approach, Israeli Minister of Housing and Construction Ariel Sharon stated: 'It would be unparalleled irresponsibility to ignore the danger of missiles, whether ground-to-ground or those dispatched from vessels or planes. But it would be more deadly and dangerous to ignore the fact that the chief

---

[15] M. Widlandski (ed.), *Can Israel Survive a Palestinian State?* (Jerusalem: Institute for Advanced Strategic and Political Studies, 1985), 10.

[16] Ibid. 11.

[17] Ibid. 14.

menace is that of the ground forces, which attack and occupy.'[18] Dore Gold, director of the US Foreign and Defence Policy Project of the Jaffee Centre for Strategic Studies, Tel Aviv University, has argued that, 'if the conventional battlefield—and not just missile attacks alone—still determines who won the war, then the conditions affecting the outcome of conventional war—from topography to strategic depth—are still critical for defining the security of a nation'.[19]

On the other side of this equation are those who see the occupied territories as a liability rather than an asset to Israeli security. Defence analyst Ze'ev Schiff has argued that Israel should reassess the value of territory to its security. He states: 'The importance of territory (in this case the West Bank) for Israel's defence cannot be dismissed, but territory does not always enhance security. Under certain conditions, like those existing in the West Bank and Gaza Strip, the risks posed by additional territory are greater than the benefits they accord.'[20]

According to this argument, the best way to enhance Israeli security may now be through a land-for-peace settlement that removes some of the sources of tension in the area and creates a more stable security environment. A continued Israeli presence along the Jordan River, and early-warning installations on the ridge line in the West Bank, could provide the means for verifying the demilitarization of the West Bank, and assuring against surprise attack from the east. Increasingly, satellite reconnaissance can improve Israel's ability to detect hostile military action as well. For the Gaza Strip, an Israeli naval presence off the coast and army presence on land would be sufficient to deny access to hostile forces, assuming Egypt was still at peace with Israel and could patrol the region's southern border.[21] While not providing the Palestinians with complete sovereignty or providing Israel with complete security, this approach, according to its advocates,

[18] A. Sharon, 'It's Time for the Chief of General Staff to Speak Out', *Jerusalem Post International Edition*, 11 May 1991, p. 7.

[19] D. Gold, 'Territory vs. Missiles: The Great Debate', *Jerusalem Post*, 13 Apr. 1991. See also D. Gold, 'Israel and the Gulf Crisis: Changing Security Requirements on the Eastern Front', *Policy Focus* (Washington Institute for Near East Policy, Dec. 1990).

[20] Ze'ev Schiff, 'Israel after the War', *Foreign Affairs*, 70/2 (spring 1991), 29.

[21] E. Sneh, 'Current Debate: How Dangerous would a Palestinian State Be? We Can Live With It', *Tikkun* (July–Aug. 1990), 65.

would protect both parties sufficiently and would be in the spirit of territorial compromise necessary for a peace agreement.

Walid Khalidi, a Research Fellow at the Harvard Center for International Affairs, and closely affiliated to the Palestinian National Council, lends support to some of the above arguments. He wrote in 1978 that 'there is no reason why the concept of Palestinian sovereignty should not accommodate provisions designed to allay legitimate fears of neighbours on a reasonable and preferably reciprocal basis'.[22] Khalidi argued further that a Palestinian state would pose little threat to Israel even if it had a small military because it would be geographically separated from most Arab states, and almost completely surrounded by Israeli territory. According to Khalidi, a future Palestinian leadership would have 'few illusions about the efficacy of revolutionary armed struggle in any direct confrontation with Israel'.[23]

A more pessimistic view, expressed by former deputy mayor of Jerusalem, Meron Benvenisti, has argued that successive Israeli governments have engaged in so much de facto annexation of the territories for military, political, and economic reasons that it is not now unrealistic to separate them physically from Israel proper.

The process of economic integration [he wrote] has long since been accomplished; infrastructure grids have been linked (roads, electricity, water, communication), administrative systems have been unified; social stratification has become institutionalized and political relationships have settled into well-established patterns.[24]

This calls into question the economic viability of an independent Palestinian state in the West Bank.[25]

Other factors complicating an Israeli withdrawal from the territories include the physical separation of the West Bank and the Gaza Strip, and the security problems presented by Israel's Arab neighbours, especially Syria, Iraq, Jordan, and Saudi Arabia. Some have proposed that Israel and its neighbours pursue a pro-

[22] W. Khalidi, 'Thinking the Unthinkable: A Sovereign Palestinian State', Foreign Affairs, 56/4 (July 1978), 701.
[23] Ibid. 713.
[24] M. Benvenisti, 1987 Report: Demographic, Economic, Legal, Social and Political Developments in the West Bank (Jerusalem: Jerusalem Post Press, 1987), 70.
[25] M. Benvenisti, The West Bank and Gaza Atlas (Jerusalem: Jerusalem Post Press, 1988), 43.

gramme of mutual security, based on a confederation among Israel, Jordan, and a Palestinian state. In addition to the confederation, the withdrawal of Israel from the West Bank and Gaza Strip would be accompanied by demilitarization, the right of Israel to pursue extra-territorial activities in case of agreement violations, and a reduced Israeli military presence in areas where mutual security arrangements had been negotiated. Intelligence stations could be relocated within Israel, while ground stations would need to be located in the territories; these could be manned either by US personnel or by joint Israeli, Jordanian, and Palestinian teams.[26]

In this regard, territorial adjustments to the green line for security purposes would have to take into account the Arab populations living along the line. Israel would not be able to argue successfully for new political boundaries that forced more Arab towns and villages to live under Israeli jurisdiction. However, there are a number of areas in the West Bank that offer both strategic and tactical assets, but have very low Arab densities. The entire eastern third of Samaria and Judaea, for example, is almost completely devoid of Arab population.[27] At the same time, this area offers some of the best surveillance points and marshalling areas to protect against an incursion from the east across the Jordan River.

## Demilitarized Zones; Peace-keeping and Verification Forces

There is now enough experience with peace-keeping forces in the Arab—Israeli conflict to say that, in conjunction with demilitarized zones, they will be essential for a new land-for-peace agreement. They not only have great symbolic value, but are instrumental in providing surveillance and verifying compliance. Although peace-keeping forces cannot prevent military confrontation, they raise the threshold for violence and remove many sources of insecurity. In view of the success of the Multinational Force of Observers in

[26] Ze'ev Schiff, *Security for Peace: Israel's Minimal Security Requirements in Negotiations with the Palestinians* (Washington DC: Washington Institute for Near East Policy, 1989), 51–8. For a comprehensive overview of the various territorial options, see *The West Bank and Gaza: Israeli Options for Peace* (Tel Aviv: Jaffee Centre for Strategic Studies, 1989).

[27] See S. Cohen, *Israel's Defensible Borders* (Tel Aviv: Jaffee Centre for Strategic Studies, 1983), 35.

the Sinai and Israel's residual suspicions of the United Nations, it is likely that a US, or possibly US–Russian, military presence may be necessary to achieve further disengagements in the Golan Heights and to establish a buffer zone along the Jordan Valley. Working out the details of such an operation would be complicated but, as suggested, the precedents are good.

An essential component of new peace-keeping forces will be the establishment of verification regimes and surveillance systems. Experience in the Israeli–Syrian and Israeli–Egyptian cases shows that verification and surveillance measures are effective in assuring compliance and maintenance of disengagement agreements. Verification and control measures have worked in the past because both sides saw it as in their interests to meet the demands of their agreements. These measures cannot be used to create agreements; they can only ease their acceptance and promote their implementation.

### Palestinian–Israeli Security Arrangements

Israeli flexibility on the nature of the Palestinian regime to be carved out of the occupied territories will vary according to whether agreement is reached with Jordan, Syria, and other Arab countries on mutual security issues. Without a clear guarantee that its external-security problem is resolved, no Israeli government is likely to accept anything more than limited autonomy for the Palestinians, with Israel controlling all external relations and retaining a military presence in key strategic regions of the occupied territories.

Concerning Israel's internal-security problems, one of the most difficult questions to resolve concerns the internal security of a new Palestinian regime and how this will impact on Israel itself. There has been very little written about this topic, in part because the nature of the security regime will itself be dependent on the type of political settlement reached between Israel and the Palestinians. For example, will the new borders demarcating Israel and Palestine be closed or open? What institutions within the Palestinian area will be given power to enforce internal security and what weapons will they be permitted to have? Will Israel have the right to liaise with the Palestinian security forces, or to engage in hot pursuit of terrorists? Which Palestinians will be freed from Israeli jails, and which members of extremist Palestinian diaspora groups will be permitted to return? These questions are both complicated and highly sensitive, but must be answered before a realistic appraisal of the security environment is possible.

Imagine, for instance, a situation in which Palestinian leaders acceptable to Israel have negotiated autonomy at a peace conference, but the outcome is bitterly opposed by Palestinian extremists both within the territories and within the diaspora. How does one prevent radical PLO members from returning to Palestine, or Islamic fundamentalists from launching an armed uprising? The dilemma is clear: if the terms of autonomy or statehood require the Palestinians to lower their enforcement capabilities to a small militia that poses no military threat to Israel, it may not be strong enough to cope with internal violence. Yet, the alternatives—a strong, well-armed militia or formal arrangements to permit Israeli intervention—are equally difficult to imagine.

The internal-security problem is directly related to the type of border arrangements negotiated between the parties. There are strong arguments expressed by some Israelis and Palestinians that, ultimately, the borders between Israel and the Palestine regime must be open. In this way, Israeli access to Jewish settlements in the new Palestine region will be assured; similarly, however, it will also give Palestinians the ability to move freely into Israel, raising the highly sensitive question of the Palestinian right of return to all locations in former Palestine, including Jaffa and Haifa. Open borders also make sense for use of the common infrastructure—water, utilities, transportation—that has been developed over the past twenty-five years.

Others, however, argue that open borders will not work, in part for the reasons addressed above, but also because Palestinians and Israelis do not want to live together in such close proximity. A controlled border is necessary for political, economic, and security reasons. Once a border of this type is envisaged, the question of geography and security becomes paramount. A geographical divorce between Israel and the Palestinian regime virtually guarantees that Israel will insist on major revisions to the pre-1967 borders and accept nothing less than complete control of the

Jordan Valley and access to the high ground along the Judaean–Samarian Hills.[28]

## Non-Conventional-Weapons Agreements

Once there is a peace between Israel and its neighbours, it will become much more practical to discuss regimes to eliminate weapons of mass destruction and their means of production. No matter what happens in the multinational fora to restrict these categories of weapons, a regional agreement will hinge on former President Ronald Reagan's phrase: 'trust but verify.'

Trust between Israel and the Arabs will not be easy to achieve. Furthermore, since the regional conflict involving Israel has an important religious component, the potential hostility of non-Arab Muslim countries such as Iran and Pakistan cannot be ignored. For instance, it is difficult to imagine an Arab–Israeli agreement to limit nuclear weapons that ignored the nuclear capabilities of Iran and Pakistan. Yet, the political circumstances under which both of these countries might be brought into a Middle Eastern nuclear-free zone will be subject to different considerations from those that will contribute to an Arab–Israeli peace.

In October 1991 US intelligence officials concluded that Iran was aggressively seeking a nuclear weapon, and that China had provided some of the fissile material for the production of such a weapon.[29] Furthermore, in addition to co-operation with China, evidence has emerged of Iran's unsuccessful attempts to obtain nuclear-related technology from Brazil.[30]

The process of identifying Iran's efforts to acquire nuclear material is further complicated by the fact that the particular technology Iran is pursuing can be used for peaceful purposes. According to one US government analyst,

90 per cent of what Iran is seeking from foreign suppliers can be used equally for nuclear weapons and civilian power, providing a ready cover

[28] See M. A. Heller and S. Nusseibeh, *No Trumpets, No Drums: A Two-State Settlement of the Israeli–Palestinian Conflict* (New York: Hill and Wang, 1991).

[29] See J. Smith, 'Officials Say Iran is Seeking Nuclear Weapons Capability', *Washington Post*, 30 Oct. 1991.

[30] Ibid.

for the weapons-related work...the Iranian shopping list includes nuclear fuel, equipment for handling and processing fissile materials, and nuclear reactors to replace those destroyed in the 1980–88 war with Iraq.[31]

Another complicating factor is that the ultimate justification for Israel's nuclear-weapons programme, including pressures to upgrade it and build a thermonuclear weapon, is the vast asymmetry in geography, population, and resources in favour of the Arab and Muslim countries. Thus, agreements on weapons of mass destruction cannot, in the last resort, be decoupled from efforts to put restrictions on conventional weapons, defence-spending, and weapons production.

Another complication concerns verification. A regional regime that authorizes the level of intrusiveness necessary to detect covert activity in the chemical, biological, and nuclear fields will be extremely difficult to negotiate, since so many parties are involved. No matter what ground rules are established in Geneva or in Vienna, the trust required for Israel, the Arabs, and other Muslim countries to agree to full inspection of one another's facilities assumes, not just peace, but a lasting and established peace.

## Conventional-Force Reductions

Since conventional-force reductions will ultimately be linked to limits on weapons of mass destruction, the model of force reductions currently negotiated in Europe may be of limited value as an indicator of how such reductions would be implemented in the Middle East. Clearly, there are some conventional-force capabilities that will be perceived by the parties as more threatening than others. If mutually acceptable equations for military balance were negotiated, so much the better. However, each country in the region faces different types of threats. Deciding what is a realistic force level for Saudi Arabia *vis-à-vis* Israel as distinct from Saudi Arabia *vis-à-vis* Iran and Iraq, for example, will require complicated technical negotiations. Nevertheless, arms-control limitations for conventional weapons, if paralleled by supplier agreements on weapons sales, will be essential if constraints on weapons of mass destruction are to be implemented.

[31] Ibid.

### Inspection of Arms Production

The Arabs will pay particular attention to the need to inspect arms-production facilities, in view of the very sophisticated nature of Israel's defence industry and its close co-operation with the United States. This issue will become increasingly complex, as so many of the components of modern military forces come from dual-use technologies. Israel's future economy is dependent on a high level of technological development to utilize or employ all the skilled labour that has come from the Soviet Union. Thus, to monitor what countries like Israel are producing in the military, as distinct from the non-military, field will be difficult, and can only be undertaken if mutual trust and co-operation in many other areas are forthcoming.

### Regional Arms-Supply Agreements

If an Arab–Israeli peace exists and basic security problems have been resolved, it is realistic to discuss combined supplier and regional power agreements on arms limitations. There will be strong economic motives to hold down levels of defence spending after peace has been achieved. In the last resort, this may be the most promising development for regional co-operation between suppliers and recipients.

### CONCLUSION

The above options are all feasible following a peace settlement. However, there can be no major reductions in force levels until the parties agree to them. It may be possible to design cosmetic arms-control agreements immediately after a peace treaty is signed, but the serious cuts will have to wait until peace has had a chance to work.

Nevertheless, in the long run it must be in the interests of all Middle Eastern countries to subscribe to the concept of a zone free of mass-destruction weapons, and to put limits on conventional forces. The fundamental asymmetries of the Arab–Israeli conflict make it difficult to conceive of stable deterrence based on threats of mutual terror.

# 12

# Controlling the Arms Trade:
# Prospects for the Future

## STEPHANIE NEUMAN

ARMS-CONTROL fever is in the air. Limiting the transfer of conventional weapons, once thought to be a utopian dream, is now discussed as a viable policy option by policy-makers and the media. The United States and the Soviet Union are reducing the number of strategic and conventional weapons in their arsenals. Two months after the February 1991 Gulf War ceasefire, President Bush announced his Middle Eastern arms-control initiative for restraints on transfers of conventional arms to the region and a freeze on surface-to-surface-missile sales. In October 1991 the big five major exporters—the United States, the United Kingdom, France, the Soviet Union, and China[1]—agreed to exchange information and limit major weapons transfers to the Middle East. The following December, the General Assembly of the United Nations voted 150–0 in favour of an arms-transfer register to collect and publish information on arms transfers in order to achieve transparency, promote restraint in the arms trade, and build confidence among states. President Bush,[2] Prime Minister Mulroney, President Mitterrand, and other world leaders have proposed various schemes further to limit international arms sales. And, on the national level, France and Germany have both tightened legislative loopholes regarding the export of weapons of mass destruction.[3] Several other states have initiated more restric-

[1] The 'big five' account for 85 per cent of the arms deliveries to the Middle East.
[2] The Bush Administration has articulated several foreign-policy goals which entail arms-transfer restraint. In addition to limiting the level of armament in the Middle East, they include preventing the proliferation of nuclear, chemical, and biological weapons, and controls on the transfer of high technology.
[3] In 1992, for example, embarrassed by disclosures that German companies had helped Iraq and Libya build chemical plants and had sent equipment to Iraq that could be used to build nuclear weapons, the German *Bundestag* created a new

tive domestic arms-export regulations: Italy has frozen arms exports, and, since 1989, Austria, Belgium, Finland, Norway, Sweden, and Switzerland have stiffened their export policies.[4]

Why has world opinion shifted on this issue? What constellation of events has brought about this dramatic change?

### THE 1991 GULF WAR: MORAL OUTRAGE AND ENLIGHTENED SELF-INTEREST

The Gulf War was a watershed, mobilizing world public opinion in favour of arms-transfer restraints. Prompted by feelings of moral outrage at the behaviour of Saddam Hussein, the war epitomized, particularly to the developed world, the dangers of weapons proliferation to areas of high tension and to aggressive countries, such as Iraq. Before the invasion of Kuwait, the Middle East had been the leading importer of military equipment—$84 billion or one-third of the world's total between 1985 and 1989—much of it advanced technology. Iraq alone accounted for over one-quarter of the regional total.

For the militaries of the United States, the former Soviet Union, and the European allies, it was also a case of enlightened self-interest. The war alerted them to the potential threat their own troops now faced in theatres of conflict once considered peripheral to the central war zone. When it ended, the US military voiced strong misgivings about the diffusion of advanced and intermediate conventional weapons—weapons that, in its view, increase US vulnerability in regional warfare. Included were: submarines;[5]

---

Government agency to monitor exports. It also approved bills that will allow customs police to tap telephones and intercept mail to stop illegal shipments of arms. ('Germany Acts to Curb Arms Exports', *New York Times*, 14 Jan. 1992; 'Germany to Allow Taps in Arms-Related Exports', *Washington Post*, 24 Jan. 1992.)

[4] A. Karp, 'A Farewell to the Arms Trade', unpublished Ms (8 May 1991, Stockholm), 24.

[5] Some 41 countries collectively have 393 submarines, and 19 countries are currently building them or have recently done so. Taiwan, South Korea, and possibly Chile and Canada are preparing to build submarines in the future. For a detailed analysis of the proliferation of submarine technology, see Statement of Rear Admiral Thomas A. Brooks, USN, Director of Naval Intelligence, Before the Seapower, Strategic, and Critical Materials Subcommittee of the House Armed Services Committee on Intelligence Issues, 7 Mar. 1991, 59–64 (hereafter referred to as Brooks Statement).

stealth, and low and very low observable technology;[6] reconnaissance and space systems;[7] high-speed computers,[8] and electronic-warfare technology.[9]

Compounding the general feeling of disquiet is the spread of weapons of mass destruction along with their production facilities. The war which saw Saddam Hussein rain Scud missiles on the allies and Israel and threaten to use chemical weapons against them was a sobering experience. In the West, the outrage has been intensified by the knowledge that Iraq purchased Western arms and technologies illicitly, and that the Iraqi government lied regarding its nuclear-production capabilities and intentions, in spite of the West's assistance and support during the Iran–Iraq war.

## US PRIMACY AND THE DEMISE OF THE COLD WAR

Other factors have also contributed to the growing enthusiasm for arms controls. The reduced tension between the United States and the former Soviet Union has prompted both governments to rethink their policies towards the Third World. During the Cold War period each tended to construe regional conflicts in terms of a zero sum game, whereby the enlistment and support of proxies were considered necessary to halt the political or territorial advancement of the other. A gain by one side's proxy was deemed

[6] The fear is that this technology will allow countries to modify their equipment to reduce signatures and vulnerability to attack (Brooks Statement, 69).

[7] During the 1990s, selected countries that have embarked upon indigenous space-based reconnaissance development will be able to achieve electro-optical imagery resolutions of less than one metre. France, India, Japan, the United States, and China have imagery satellites. Countries that could have imagery satellites by the year 2000 include: Canada, Germany, Israel, Italy, Pakistan, South Africa, South Korea, Spain, and Taiwan. Argentina and Brazil could have them sometime after 2000.

Apart from the European Space Agency and the superpowers, four countries (India, Israel, Japan, and China) now have the capability to launch satellites; Pakistan, South Africa, Taiwan, and others are developing the capability (Brooks Statement, 70–1).

[8] 'New Curbs on Exports Are Sought', *New York Times*, 11 Sept. 1991, D1, D28.

[9] Many of these systems are already being exported; problems in integrating various subsystems/components are delaying effective operational use of electronic-warfare technology systems in most countries; as computers continue to be exported, however, these systems will become more user friendly (Brooks Statement, 75).

a loss for the other superpower. Today, both the United States and the troubled newly independent republics of the former Soviet Union are more likely to perceive Third World conflicts as internal or civil wars. Increasingly, military or economic involvement is considered an unnecessary and insupportable expense, as the 1991 US–Soviet agreement to end arms sales to rival sides in the Afghan War portended.[10] Since the former Soviet Union's withdrawal from Afghanistan in February 1989, continued US support for the insurgents has become harder to defend in Congress and to the American people. For the independent republics, removing irritants in their relations with the United States is seen as a way of earning vital economic assistance.

Closely associated with the diminished tension between the United States and the former Soviet Union is the economic and political disarray in the newly independent republics. Seeking economic and technological assistance, and sorely in need of a tranquil international environment in order to stabilize their respective regimes, the republics of the former Soviet Union are eager to co-operate with the United States. They are unlikely, on this score alone, to challenge US arms-transfer initiatives. To date, control of the arms trade has been identified by both former Soviet President Gorbachev, and Russian President Yeltsin, as an issue of mutual concern.[11]

[10] 'US and Soviets to End Arms Sales to Afghan Rivals', *New York Times*, 14 Sept. 1991, A1, A4.

[11] In Gorbachev's and Yeltsin's 'Joint Statement on Non-Proliferation' signed in 1990 with the United States, the Soviet Union accepted the provisions of the MTCR designed to halt the proliferation of missile technology. Foreign Minister Eduard Shevardnadze, in a letter to the UN Secretary General, 15 Apr. 1990, suggested limiting international sales and supplies of conventional weapons as 'a means of building a new model of security...' (letter to the UN Secretary General, *Izvestia*, 15 Aug. 1990, cited in SIPRI, *Yearbook 1991*, 220). Sergei Grigoryev, deputy to President Gorbachev's chief spokesman, stated that developed nations should agree to limit arms sales to 'dangerous' countries (summarized from wire copy, *Current News*, 19 Feb. 1991). In 1991 the Soviet Union welcomed a US proposal to ban chemical weapons (summarized from wire copy, *Current News*, 24 May 1991). President Yeltsin, in a 27 Jan. 1992 statement to the UN Secretary-General declared that 'Russia considers itself the successor to the USSR with regard to the responsibility for carrying out international obligations'. He reaffirmed Russia's obligations under the Treaty on the Non-Proliferation of Nuclear Weapons, its support for the arms trade guidelines approved in London, October 1991, for greater transparency in the military sphere, and for a global convention on the prohibition of chemical and biological weapons (UN General Assembly, Security Council, A/47/77; S/23486, 28 Jan. 1992).

The 1991 Gulf War also demonstrated incontrovertibly what past events in the former Soviet Union implied—that the United States is the predominant political and military power in the international system. In 1991 a US official for the first time publicly declared the United States to be the sole world superpower. Itemizing measures of economic, political, military, and cultural power, Robert Gates, then Deputy National Security Adviser, declared: 'We have no challengers.'[12] Once universally perceived as bipolar, the international system is now regarded as unipolar by a growing number of analysts.[13] Why this is so is neatly summarized by Catherine Kelleher:

The US is uniquely capable of leading well into the next century because it has been, and will continue to be, suited for co-optive leadership.[14] The United States has the resources and universal culture that appeal to a broad faction of the global community; the system of alliances it leads and the informal coalitions it has built now constitute the most effective global management system. No other power or group of states will be able in the foreseeable future to command or co-opt this system.[15]

The significance of all this for arms control and arms transfers is far-reaching. As international tensions have diminished, tighter arms-transfer restrictions have become more attractive to US policy-makers. Given its predominance in the system, US preferences on this issue are likely to receive a favourable response from states now more dependent than ever on US political and economic largess.

---

[12] 'US Declares itself the Sole Superpower', *Washington Times*, 8 May 1991, p. 1.

[13] See, e.g., C. Krauthammer, 'The Unipolar Moment', *Foreign Affairs: America and the World, 1990/91*, 70/1 (1991), 23–33; L. Freedman, 'The Gulf War and the New World Order', *Survival*, 33/3 (May–June 1991), 195–209.

[14] 'Co-optive leadership' refers to Joseph Nye's definition of power. He distinguishes between two types in international relations; command and co-optive power. Command power is 'the ability to change what others do', and co-optive power is 'the ability to shape what others want'. Command power is associated with coercion or inducement, utilizing such 'hard resources' as military or economic sanctions. Co-optive power is associated with use of 'soft power resources', such as the attractiveness of one's culture or ideology (J. Nye, *Bound to Lead: The Changing Nature of American Power* (New York: Basic Books, 1990), 267).

[15] C. M. Kelleher, 'The Changing Currency of Power: Paper I, The Future Nature of US Influence in Western Europe and North-East Asia', *America's Role in a Changing World*, (Adelphi Papers, 256; winter 1990–1), 23–36, at pp. 28–9.

## STRUCTURAL ARMS CONTROL

Economic factors are also enhancing the prospects for global arms-transfer restraint. World recession combined with changes in the world political climate has created a situation in which the demand for military equipment is in decline. Declining demand has set into motion a cycle whereby shorter production runs raise the unit cost of military equipment,[16] which, in turn, places an additional burden on already shrinking budgetary resources, further decreasing the number of weapons the military can afford to buy. As a result, military industries everywhere are seeking financial relief through bankruptcy, production cutbacks, conversion, mergers, or a host of other economizing measures. As conditions continue to deteriorate, more countries may see it in their economic and political interest to curtail not only their own procurements but, through arms-control agreements, those of their enemies as well. Israel, for example, is now taking a new look at internationally organized controls. 'This is a change in attitude,' observed a Foreign Ministry official. 'We preferred to stay out of international organs agreements because we felt they were politicized and stacked against us. We want to get involved now, if for no other reason than to see that others don't cheat.'[17]

There are many indicators of structural arms control. Worldwide arms imports have declined dramatically in recent years, dropping 28 per cent between 1987 and 1989. Imports fell most precipitously, however, in the Third World (33 per cent) and former Warsaw Pact countries (36 per cent). In the Third World, the Middle East's arms imports plunged almost 50 per cent, in contrast to Latin America (37 per cent) and Africa and east Asia (34 per cent). Only south Asia's military procurements increased in dollar value.[18]

World-wide military production, too, is expected to drop—

---

[16] François Heisbourg found that the price indexes of arms, aerospace, and shipbuilding in comparison to other categories of manufactured goods were 26–90 percentage points higher (F. Heisbourg, 'Public Policy and the Creation of a European Arms Market', in P. Creasy and S. May (eds.), *European Armaments Market and Procurement Cooperation* (London: Macmillan, 1988), 60–88, at p. 61).

[17] 'Israel Battling Arms Sales to Foes', *Los Angeles Times*, 2 Jan. 1992, p. 10.

[18] US ACDA, *World Military Expenditures and Arms Transfers, 1990* (Washington DC: US Government Printing Office, 1991).

some estimate by as much as 20–30 per cent in the 1990s.[19] For arms producers in general, falling defence budgets and contracted demand have had serious negative consequences for their exports and military industrial infrastructure. In an ever-reiterative downward spiral, small and medium-size states' ability to produce sophisticated weapons is dwindling, as their dependence upon the more advanced military-production capabilities of others, particularly those of the United States, is growing.

## Western Europe

In Western Europe not only is defence production declining for all the above reasons, but large numbers of domestic as well as collaborative programmes are being cancelled, particularly for next-generation platforms, such as tanks and planes. As one analyst concludes, 'It no longer seems possible to deny that in the short to medium term, aggregate defense production in Europe is declining rapidly and that a major restructuring is underway.'[20]

Individual European states do not have the military market that allows for the economies of scale and the spreading-out of research-and-development costs enjoyed by US industries. The situation is exacerbated by the rapidly escalating cost of production, the speed of technological change, and the attendant altered threat environment in the shape of deep-strike aircraft and missiles which demand countermeasures in the realm of Extended Air Defence capabilities.[21] As former French Defence Minister Jean-Pierre Chevenement divulged in January 1991, France can no longer afford to manufacture 84 per cent of its weapons.[22]

These economic realities suggest that the Europeans will remain dependent on US high-technology inputs into their defence industries, and perhaps on off-the-shelf purchases of major

[19] K. Krause, 'Trends in the Production and Trade of Conventional Weapons', paper prepared for the conference on 'The Supply-Side Control of Weapons Proliferation', Canadian Institute for International Peace and Security, Ottawa, Ontario, Canada, June 1991, pp. 6–7.

[20] E. J. Laurance, 'A Model and Some Preliminary Indications', paper prepared for the annual meeting of the International Studies Association, Vancouver, Mar. 1991, p. 8.

[21] Heisbourg, 'Public Policy', 60.

[22] Interview with Gerard Renon, 'One on One', *Defense News*, 27 May 1991, p. 38.

weapon systems as well. If the past augurs the future, the latest published arms-export data support this view. Between 1985 and 1989 the United States exported (in current $) $22 billion in defence equipment to Europe and imported $6 billion, roughly a 4–1 trade advantage in favour of the United States.[23] Even France, which in the past made the greatest effort to maintain an independent military-production capability, is considering the purchase of foreign equipment.[24] France's former Under-Secretary of Defence declared in May 1991: 'There are some types of equipment that can be bought off the shelf without endangering our security. And if costs are attractive we could buy foreign equipment rather than launching French industry in complex and expensive development programs.'[25] If European states do increase their military purchases from the United States, however, many of their domestically produced military goods as well as their imported major systems will be subject to US third-party transfer prohibitions, providing the United States with an important means of controlling their spread.

The prospects for achieving economies of scale and increased independence through integrated defence production is the wild card. Should Europe be successful in these efforts, as some predict, Europe's dependence upon the United States will diminish. In November 1990 twelve EC states organized Euclid to co-operate on research and development for defence to boost European technology and challenge US leadership.[26]

But there is some doubt whether Europe can compete independently with the United States. Thus far it has been unable to do so in major critical technologies. In a 1990 study, the US

[23] ACDA, *World Military Expenditures, 1990*; US DSAA, *Foreign Military Sales, Foreign Military Construction, Sales, and Military Assistance Facts*, as of 30 Sept. 1990 (published by Data Management Division, Comptroller, DSAA).

[24] France imported $966 million worth of armaments from the United States between 1985 and 1989 or about 4 per cent of Europe's total (DSAA, *Foreign Military Sales*).

[25] Interview with Gerard Renon, 'One on One', 38.

[26] 'Europe Searches for own Voice in Future International Conflicts', *Aviation Week and Space Technology*, 24 Dec. 1990, pp. 37, 40. Euclid members are: Belgium, Germany, Netherlands, Spain, Denmark, Greece, Norway, Turkey, France, Italy, Portugal, and the United Kingdom. It operates under the jurisdiction of the Independent European Program Group, which was founded in 1976 to facilitate closer European armament co-operation and to improve European competitiveness with the United States.

Defense Department found that, for twenty-two critical technologies, NATO Europe was on a par with the United States in seven—about one-third—but, unlike Japan, was not ahead of the United States in any category.[27] Analysing the strengths and weaknesses of Western Europe's military industrial capability, sector by sector, another study concludes: 'Western Europe tends to be well established in older technologies—heavy mechanical engineering and hydrodynamics, for example—and fully on a par with America in these; but it occupies a weaker position in the newer and emerging technologies, especially in the vital field of electronics.'[28]

Furthermore, experience so far has been that collaborative projects are inefficient and costly. With the huge gap in research-and-development expenditures (the United States spends four times more than all of Europe combined on military research and development) it would appear that, even with co-operative efforts, unless Europe decides to invest significantly more resources in military research and development, it will not be competitive enough either to produce what it needs independently or to increase its exports substantially.[29]

Nationalism has been another obstacle to European collaboration in the past, and one that may persist in the future. Along with a general feeling of loss of sovereignty, there exist differences in doctrine, threat perceptions, and defence requirements which make co-operation on military production and procurement difficult. There is also concern about lost jobs, and the response of the electorate to them. As a result, each country has tended to believe that it is financing other countries' programmes and interests in co-operative ventures, at the expense of its own. And then there are other more subtle types of competition, as exemplified by the Euclid project, where, after some dissension, it was agreed to spell Euclid the English rather than the French way.[30]

[27] G. Leopold, 'US Faces Tough Competition in Critical Technologies', *Defense News*, 16 July 1990.
[28] P. Bates, 'Defence Technology in Western Europe', in J. D. Drown, C. Drown, and K. Campbell (eds.), *A Single European Arms Industry? European Defence Industries in the 1990s* (London: Brassey's, 1990), 105.
[29] For a discussion of the rising relative share of research and development in major weapons programmes and the implications for Europe, see Heisbourg, 'Public Policy', 62–6.
[30] 'Europe Searches for own Voice', 37, 40.

The best proof of the pudding is in the eating, and, since 1957, when the first Franco-German co-operative agreement was initiated, Europe has partaken only sparingly of military industrial collaboration. Co-operative-procurement ventures represent a fraction of European weapons programmes: 15 per cent for the United Kingdom, and no more than 20 per cent of French conventional-arms procurement.[31]

In sum, given depressed markets elsewhere, Europe is eager to maintain access to vital US civilian and military markets[32]—and is therefore sensitive to US arms-control preferences and pressures. If the price of access is continued procurement from US defence industries and restrictions on arms transfers, then the likelihood of maintaining an independent European production capability and export policy becomes more remote. Some interpret Europe's support of Arab−Israeli negotiations and arms-transfer restrictions to the Middle East as denoting a more flexible and co-operative attitude.[33]

## The Former Soviet Union

It is not necessary to detail the economic chaos unfolding in the former Soviet Union. Conditions in the military sector have been equally tumultuous, both in the Soviet Union before its disintegration and in the Russian Republic, which inherited over 80 per cent of the former Soviet Union's aerospace and defence industries.[34]

[31] Heisbourg, 'Public Policy', 60. The percentage may be even lower for the United Kingdom. *Jane's* reports that 90 per cent of the equipment the Ministry of Defence buys come from the UK (*Jane's Defence Weekly*, 7 Dec. 1991, p. 1081).
[32] The US economy and military market are the largest in the world. The US GNP ($5.2 trillion) is almost twice that of its nearest competitor, Japan (ACDA, *World Military Expenditures, 1990*). In the military sector, even after budget-cutting, the US procurement allocation of $64.3 billion for FY 1992 (down by 4 per cent from 1991) is still the largest in the world (Department of Defense Appropriations Act, 1992, Public Law 102−172, 26 Nov. 1991; Department of Defense Appropriations Act, 1991, Public Law 101−511, 5 Nov. 1990).
[33] Mark Kramer argues that Europe was less co-operative in the past, complicating past US efforts in the region by providing alternate sources of arms, aid, and political support. He cites Iraq's missile and nuclear programmes as only one example (M. Kramer, 'Army Won't Be a Loose Cannon', *Los Angeles Times*, 4 Dec. 1991, B5).
[34] Together, the Russian Republic and the Ukrainian Republic account for 90 per cent of the former Soviet Union's military industrial sector ('Industry Nears Independence', *Jane's Defence Weekly*, 23 Nov. 1991, p. 995; 'Drastic Cuts in Weapons Orders Underway', *Aviation Week and Space Technology*, 27 Jan. 1992, p. 34).

Defence-spending in the former Soviet Union is estimated to have declined about 6 per cent in 1989 and 1990,[35] causing severe production and research cutbacks throughout the industry. Military orders to Soviet defence plants fell by 21 per cent in 1991 from the previous year, and in January 1992 the Russian parliament approved a budget that slashed arms purchases to about one-seventh of the 1991 level. Research-and-development funding may fall by as much as 30 per cent.[36] By 1992 factories were complaining that military contracts had been reduced to practically nothing.[37] Supplies, particularly those needed for civilian manufactures, had become increasingly difficult to obtain, and, where production was still continuing, factories were drawing on existing stocks of material. Unless government policy shifts, some analysts believe that, once these stockpiles are exhausted, these factories will close.[38]

Arms exports from the former Soviet Union have plummeted in recent years. In the past, almost 90 per cent went to the Third World. Beginning in 1988, however, the dollar value of deliveries to these countries began a downward slide, slipping 13 per cent between 1988 and 1989 and 28 per cent between 1989 and 1990.[39] Dramatic declines were recorded in tonnage as well.[40]

The defeat of Soviet military systems manned by Iraqis during the 1991 Gulf War has done little to encourage sales of Soviet equipment. Moreover, if the Russian government adheres to its declared policy of conducting all foreign trade in hard currency and commercially pricing much of its military equipment, sales are likely to contract further. The comparative advantage of military equipment from the former Soviet Union—low cost at generous terms—will evaporate, making it even less attractive to dis-

[35] Interview with Uri Ryzhov, Chairman, Supreme Soviet Science Committee in 'One on One', *Defense News*, 22 July 1991, p. 62.

[36] 'Drastic Cuts in Weapons Orders Under Way in Former Soviet Union', *Aviation Week and Space Technology*, 27 Jan. 1992, p. 34; 'Wire News Highlights', *Current News*, 24 Jan. 1992, p. 16; "Russia to Cut Arms Orders by 85 pc This Year', *Daily Telegraph*, 16 Jan. 1992, p. 8.

[37] 'Arms factory can make bricks, but, Russia asks, is that smart?', *New York Times*, 24 Feb. 1992, A1, A10.

[38] 'Drastic Cuts in Weapons Orders Under Way', 34.

[39] R. F. Grimmett, *Conventional Arms Transfers to the Third World, 1983–1990* (Washington DC: Congressional Research Service, 2 Aug. 1991), Table 2A, p. 59.

[40] Brooks Statement, 38.

affected, impoverished, and indebted Third World and Eastern European customers.[41]

Unless the Russian Republic is able to resolve its internal problems and maintain a sophisticated research-and-development and industrial infrastructure over the long term, its role in the world's arms-production and transfer system is likely to deteriorate further.

## The Third World

A similar state of affairs exists in the Third World. By 1989 imports of weapons had declined 31 per cent from 1987 levels and exports had declined 29 per cent.[42] For many Third World producers, whose defence industries depended upon the Iran–Iraq War for sales, reduced demand has spelt disaster. Brazil is a typical example. Its exports fell by 93 per cent between 1987 and 1989.[43] The embargo on Iraq after it invaded Kuwait eliminated Brazil's major market for Astros II rocket launchers and missiles and Urutu armoured vehicles. Because exports are vital to the health of Brazil's arms industries—some 90 per cent of production is exported[44]—many of these industries were thrown into or near bankruptcy.[45] For the foreseeable future, then, the dearth

[41] Soviet deliveries to India were down in 1990 in comparison to previous years, because of Indian budgetary constraints, stricter Soviet pricing policies, and contract completions. The hard currency policy is likely further to diminish Soviet military deliveries to India, since, in the past, India has paid in rupees. These same factors have already caused problems with China's proposed purchase of an initial batch of MiG-29s and Su-27s (*Milavnews*, NL-354, Apr. 1991). For further discussion of Third World–former Soviet Union arms transfer relations, see S. K. Purcell, 'The US and Regional Conflicts, *America's Role in a Changing World*, (Adelphi Papers, 256; winter 1990–1), 73–84, at p. 75. Some Eastern European countries, too, have expressed their desire to reduce arms purchases from the former Soviet Union. See SIPRI, *YEARBOOK, 1991*, 213, regarding the unsuccessful attempt of Czechoslovakia and the former GDR to cancel aircraft deliveries from the former Soviet Union.

[42] ACDA, *World Military Expenditures, 1990*.

[43] Ibid. According to SIPRI, *YEARBOOK, 1991*, Brazil's exports continued to drop in 1990, recording a 95 per cent decline between 1987 and 1990 (p. 198).

[44] R. Matthews, 'The Neutrals as Gunrunners', *Orbis* (winter 1991), 41–52, at p. 45.

[45] Two of Brazil's three major arms-producing industries faced bankruptcy in 1992 (testimony of Rear-Admiral Sheafer, Director of Naval Intelligence, to the Seapower, Strategic, and Critical Materials Subcommittee of the US House Armed Services Committee, 5 Feb. 1992, excerpted in *Arms Sales Monitor*, 11–12 (Jan.– Feb. 1992), 5.

of customers means that Brazil is unlikely to be a major arms supplier or manufacturer of finished end-items.

Brazil's situation typifies that of other Third World producers. As demand has fallen, so have prospects for viable defence industries, particularly for the production of major weapon systems.

Non-producers, too, are dealing with the prospect of fewer procurements. Not only have their domestic procurement budgets been slashed, but sources of military assistance are dissipating. Military grant aid from the United States—a major donor—is a declining asset.[46] Even the Gulf states, which in the past helped finance the military purchases of poorer Islamic powers, are now cutting back. Saudi Arabia, for example, has suspended financial aid to the PLO, Jordan, Sudan, Yemen, 'and to some extent to Algeria and Tunisia', countries which openly sympathized with Iraq in the Gulf crisis. Even aid to friendly countries and Arab allies, like Egypt, has been reduced.[47]

In sum, we are witnessing a structural form of conventional arms control caused by economic factors which makes arms-control initiatives potentially more attractive to recipients and suppliers. In response to the deflating arms trade, some supplier governments have found it possible to initiate more restrictive arms-export policies without paying a political price domestically. Faced with declining orders, arms industries fight less fiercely for permissive export legislation, and governments find themselves freer to respond to advocates for arms control. In general, the lower the level of arms transferred, the more feasible is tighter arms-control legislation at the national and international levels.

### REALITY VERSUS EUPHORIA

But how optimistic can we be about the success of future arms-control initiatives? How likely is it that suppliers and recipients

---

[46] The total grant aid request for FY 1993 is $4.2 billion, down $500 million from 1992. About 70 per cent of this is slated for Israel ($1.8 billion) and Egypt ($1.3 billion) (*Arms Sales Monitor*, 11–12 (Jan.–Feb. 1992), 5).

[47] 'Saudis Seek to Cut Funds for Militants', *New York Times*, 1 Mar. 1992. Saudi Arabia's vast financial reserves of $40–$50 billion before the Gulf crisis were reportedly drained by the 1991 Gulf War which cost Saudi Arabia alone $65 billion.

will agree? How probable is it that they will conform to control regulations and not be tempted to cheat?

In spite of the favourable climate, our expectations for conventional-arms limitations must remain modest. Unrealistic hopes for general and complete disarmament, or for the cessation of arms transfers, are bound to be frustrated by a variety of factors that conspire to make the implementation of even limited arms-export control measures difficult.

## Geopolitical Realities

Geopolitical realities in much of the Third World provide powerful incentives to evade arms-control restrictions. In the face of competing territorial claims, disputed borders, and unsettled ethnic and religious rivalries, many Third World governments seek arms to maintain internal order, protect their country from external threats, and ensure their own survival. From the perspective of Third World leaders, arms controls are dangerous because they may or may not deprive their enemies of the wherewithal to attack, but they certainly deprive their own country of the means of defence, particularly if it does not produce its own weapons. For them, weapons, not arms controls, are the ultimate confidence builder.

In the Middle East, for example, Syria's President Hafez al-Asad's believes the US policy on arms control in the Middle East is designed 'to strip the Arabs of their weapons' and 'halt the import of arms by Arabs while allowing Israel to manufacture arms'.[48] But, from the Israeli perspective, domestic production is vital to survival in a hostile environment, and sophisticated arms exports to neighbouring Arab states, such as the sale of F-15Es to Saudi Arabia, represent an escalated security threat and must, therefore, be controlled. 'Even if the Saudis did not initially become involved in a new Israeli–Arab confrontation, the deployment of F-15s at airbases such as Tabuk—only 6 minutes flying time from Israel—could tie down Israeli squadrons needed elsewhere,' observed one Israeli defence official.[49]

For supplier states, such as the former Soviet Union and the

---

[48] 'Syrian Accuses the US of Trying to Strip Arabs of Military Power', New York Times, 13 Mar. 1992.

[49] Near East Report, 9 Mar. 1992, p. 43.

United States, tension has always existed between their desire to limit arms transfers to potential aggressors and their self-interested need to provide allies and friends with arms for their own defence. Fear of involvement in an unwanted, unsolicited war is a powerful inducement to help friends fight their own battles. The dilemma facing the United States is how to prevent larger, more aggressive states from preying on the weak without committing US soldiers to protect them? How to initiate equitable arms controls that do not leave small states or non-producers at the mercy of their more militarily capable neighbours?

Policy debates regarding arms transfers and controls often revolve around this issue. In the Middle East, for example, the smaller Gulf states, such as Kuwait, feel particularly vulnerable. A US military attaché in the region observed: 'In the next decade, Iran will become a regional superpower and Iraq will recover. If they will be aggressive or friendly, we don't know, but Kuwait will always have to live under threat.'[50] The same holds true for the other Gulf Arab states. Their plan to build up their defence to prevent anything like the Iraqi invasion from happening again involves arms imports and military training. Responding to this security dilemma, the United States simultaneously called for arms-transfer restraint to the Gulf and, between June 1991 and January 1992, transferred $5 billion worth of weapons to the Gulf states.[51]

A similar situation exists in south Asia. The United States, uncomfortable with Pakistan's human-rights record and purported nuclear programme, cut off US military aid in October 1990. But India—which, like Pakistan, refuses to sign the NPT or to agree to international inspections and safeguards—has the material and infrastructure to produce more nuclear weapons than Pakistan and supports a much larger military industrial establishment. To Pakistan, India constitutes a major security threat. Faced with this dilemma, the United States has permitted commercial sales of arms to Pakistan of about $400 million.[52] As one Indian analyst

[50] 'The Runaway Army is Back, but Standing at Ease', New York Times, 14 Jan. 1992.
[51] Near East Report, 26/10 (9 Mar. 1992), 43. See also Arms Sales Monitor, 11–12 (Jan.–Feb. 1992), 2.
[52] 'US Expected to Pressure India on Nuclear Issue', New York Times, 10 Mar. 1992.

aptly puts it, 'unless the basic issue of the causes of insecurity and vulnerability in the developing world is adequately addressed, it would be overly optimistic to expect that the arms trade, which is a very minor aspect of the overall generation and stockpiling of arms, can be controlled'.[53]

These policy considerations will not evaporate in the future. Ironically, they make even peace treaties reliant on arms transfers for success. The precedent was set by the 1979 Israeli–Egyptian peace treaty, where the United States promised and delivered to both signatories generous military assistance to allay their fear and mistrust of each other. Should current peace negotiations in the Middle East be successful, weapons transfers are bound to be a political outcome.

## Economic Realities

Economic realities also conspire to undermine arms-control initiatives, motivating many suppliers and recipients to find ways around them. For most weapon producers, as discussed above, arms exports are a means of supporting their arms industries, achieving economies of scale for equipment required by their armed forces, and earning needed foreign currency to shore up their economies. Although only 5–15 per cent of US defence production is exported,[54] other countries are more export-dependent. Brazil, as already noted, sells 90 per cent of production abroad;[55] France exports an estimated 50 per cent of the conventional weapon systems it manufactures;[56] Czechoslovakia sells 70 per cent of what it produces;[57] and, in Russia, defence items are one of the few commodities left to vend on the world market.[58] Prime Minister I. S. Silayev revealed that 70 per cent of the

[53] J. Singh, 'Control of the Arms Trade as a Contribution to Conflict Prevention', in *Building Global Security through Co-operation: Proceedings of the Thirty-Ninth Pugwash Conference on Science and World Affairs, Cambridge, Mass., 23–8 July 1989*, 345–52, at p. 351.

[54] *Defense and Economy World Report*, 2 Mar. 1987, p. 5814.

[55] See above, p. 272.

[56] Heisbourg, 'Public Policy', 66.

[57] J. Matousek, 'Czechoslovakia', in I. Anthony (ed.), *Arms Export Regulations*, (Oxford: Oxford University Press, 1991), 50–4, at p. 51.

[58] In general, the new republics, badly in need of revenues to shore up their economies, have little to export other than weapons, advanced dual-use technologies, and parts of the space programme.

industrial enterprises on Russian territory belong to the military–industrial sector and that civilian consumer goods account for only 26 per cent of the Russian industries' production.[59]

Similar economic incentives motivate many collaborative defence projects. In addition to the M-11 missile deal with Pakistan, China entered into an agreement with Syria in 1988 to develop the M-9 missile. Both countries are reported to have provided financing for the respective missiles.[60] These ventures furnish desperately needed foreign currency for China's own military industries. As a result, however, China has export obligations to these countries, and, despite its November 1991 promise to abide by the MTCR, is expected to proceed with missile sales contracted before that date. Whether it sells the missile itself or, more likely, components and technical expertise,[61] the motive is economic and the result proliferation.

Inevitably, the need for defence exports influences how states respond to arms-control initiatives, and in many cases the economic incentives to cheat or find ways around them will be very compelling.

For these reasons, agreement among supplier states on a unified arms-control policy will face serious difficulties. Within the European Community, for example, restrictions on arms exports would inevitably have negative implications for the economies of scale of any EC collaborative production project, reducing its profits and increasing costs. Export controls would also raise questions about their differential impact on the economies of EC members. France and the United Kingdom, which are responsible for two-thirds of NATO Europe's exports, for example, have been decidedly less enthusiastic about the prospect of a common export policy than Belgium or Holland, which together represent 6 per cent.[62] Only in reference to the Gulf has there been some unanimity

---

[59] I. Silayev, 'We've Already Won Back the First Half', *Ogonek*, 24 (1991), 2, and 'If there is a Strong Russia there will be a strong Union', *Mezhdunarodnaya Zhizn*, 6 (June 1991), 11, cited in S. R. Covington and J. Lough, 'Russia's Post-Revolution Challenge: Reform of the Soviet Superpower Paradigm', *Washington Quarterly*, 15/1 (winter 1992), 5–22, at p. 17.

[60] 'US to Press China to Halt Missile Sales', *Washington Post*, 11 June 1992, p. 13.

[61] 'China Said to Sell Parts for Missiles', *New York Times*, 31 Jan. 1992.

[62] These export figures, in constant dollars, cover the 1979–89 period (ACDA, *World Military Expenditures, 1990*; see also Laurance, ('Model', 30–1).

among the twelve EC members. France, Germany, Italy, and Belgium have joined Luxemburg and the Netherlands in calls for tighter export controls, although no specific measures have been agreed upon.[63]

The imperative of technological diffusion presents yet another obstacle. The modernization needs of all militaries are a strong stimulus for activity in the arms trade and, as technology improves, vast numbers of obsolete equipment are made available—filtering out into successively less developed countries. Third World states are involved in this diffusion process as well—providing other states with old equipment they no longer need or want.[64]

### The Changing Character of the Arms Trade

The changing character of the arms trade itself challenges the prospects for conventional arms control, as what to control becomes a more baffling and complex question. Falling defence budgets are prompting states to upgrade existing inventories rather than purchase new and expensive end-items. Whereas suppliers previously transferred complete weapon systems— tanks, ships, missiles, and aircraft—today the major trade is in components, spare parts, technical assistance, and production technologies. Many of these items are shipped in crates and containers, making verification problematic, subterfuge possible, and regulation more difficult.

A particularly nettlesome issue is the increased transfer of production technology. Past and existing control regimes have focused on individual systems or components in the manufacturing process, while the export of turn-key factories which can be diverted to military usage is not prohibited. This comes at a time when automated production machinery is simplifying the task of sophisticated industrial production.[65]

[63] 'EC Mins to Examine Arms Export Restrictions to Gulf Region', *Defense News*, 25 Feb. 1991.

[64] Iraq, for example, exported captured Iranian equipment to Djibouti, Sudan, Mauritania, and Lebanon (Brooks Statement, 47). Israel, too, is reported to have transferred vintage Soviet equipment captured from the Palestinians to other Third World countries.

[65] On this point, see T. Clancy and R. Seitz, 'Five Minutes Past Midnight and Welcome to the Age of Proliferation', *National Interest*, 26 (winter 1991–2), 8–10. The authors point out that digitally controlled machine tools with optical

More and more countries are acquiring the technical capability to upgrade weapons, if not to manufacture them. Iraq, for example, has extended the range of the Scud-B to produce its Al-Hussein and the Al-Abbas missiles. Aiding and abetting this trend is the burgeoning world population of technically trained people—engineers, scientists, and technicians—who form an army of 'intellectual mercenaries' able and willing to help other states acquire more sophisticated military industrial capabilities. Largely educated in the West, many of them originating from the Third World, these men and women are returning in the thousands to their own countries, or are selling their expertise to the highest bidder in others.[66] Their ranks may soon be swelled by former Soviet defence scientists and technicians encouraged to flee their homeland by the economic chaos in the republics, the liberalization of travel restrictions, and continuing cutbacks on the former Soviet military laboratories. These are developments that will be difficult to contain in the future.

## The Blurring of Conventional and Non-Conventional Weaponry

Furthermore, the barrier between conventional and non-conventional is rapidly eroding. Much attention has been given to the control of non-conventional weapons—the so-called 'weapons of mass destruction': nuclear, biological, and chemical weapons and ballistic missiles. However, new 'conventional' systems—such as sensor and communications technologies, advanced data-processing equipment mated with new delivery systems, low observable aircraft and missiles, long-range surveillance systems, and other technologies—are expected to have a revolutionary impact on the future battlefield. The power and accuracy of these new weapons are blurring the distinction between tactical

---

sensors and air bearings have revolutionized the quality of precision machining. 'The compatibility of these precision tools with computer-aided design (CAD) and computer-aided manufacture (CAM) software as well as precision robotic manipulators, substantially lowers the demand on the proficiency and skill of their operators' (p. 10).

[66] According to Jacques Gaillard, the percentage of the world's scientists and engineers residing in developing countries rose from 7.6% to 10.2% between 1970 and 1980, and by 1990 exceeded 13% (*Scientists in the Third World* (University Press of Kentucky, 1991), cited in Clancy and Seitz, 'Five Minutes Past Midnight', 4).

and strategic weapons, and between the conventional and non-conventional.[67] As one analyst claims: 'if a full scale conventional war broke out in Europe, the battlefield two weeks later ... would be, in visual terms, little different from an attack by Hiroshima class nuclear weapons.'[68]

Many of these new technologies are also small and easily transportable, have dual-use functions, and may be available on the commercial market, making it possible for countries with meagre manufacturing capabilities dramatically to increase the power, range, and lethality of older systems. According to some, these systems are of a sophistication that can radically change 'the correlation of forces among nations used to depending on the generations of technology available on the fringes of the global arms bazaar'.[69]

The concern is that the capabilities of these less glamorous systems will be ignored by adherents of arms control. Geoffrey Kemp, for example, points out that recent developments in long-range artillery rockets begin to erase the operational distinction between short-range ballistic missiles and unguided free-fall rockets.[70] In regions where distances are short and enemies proximate, such as the Middle East, these weapons can serve as efficient delivery systems for chemical weapons.[71]

Uzi Rubin also argues that 'any relatively modern strike aircraft or combat aircraft in regional (Middle East) service can carry in one sortie roughly four times more load than a single Soviet Scud

[67] Paul W. Hoag, 'Hi-Tech Armaments, Space Militarisation, and the Third World', in C. Creighton and M. Shaw (eds.), *The Sociology of War and Peace* (London: Macmillan Press, 1987), 79; T. J. Welch, 'Technology Change and Security', *Washington Quarterly* (spring 1990), 114–15.

[68] Quoted in Hoag, 'Hi-Tech Armaments, Space Militarisation and the Third World', 79.

[69] Clancy and Seitz, 'Five Minutes Past Midnight', 8.

[70] G. Kemp, *The Control of the Middle East Arms Race* (Washington DC: Carnegie Endowment for International Peace, 1991), 197. Long-range artillery rockets in production or under development are: Egypt's Sakr 80—80-km range; Iran's Oghab—40-km range; Israel's MAR-350—90-km range.

[71] The Iraqi 131-foot long-range artillery piece—the so-called supergun—is a case in point. This enormous gun, with a 1,000-mm calibre gun tube which was shipped as oil-piping, once operational, would have been able to fire conventional and non-conventional munitions at ranges of approximately 1,600 km. 'Long range artillery over short distances can offer a less expensive alternative to strategic bombers or missiles. Moreover, as technological developments in rocket-driven artillery shells continue, the barrier between long-range artillery and short-range tactical missiles is eroding' (Kemp, *Control of the Middle East Arms Race*, 200).

ballistic missile. . . . It can be seen, then, that aircraft are by far the more significant and lethal platforms for mass destruction attacks'.[72] In some parts of the world, such as the Middle East, the conventional has become non-conventional, and arms-control regimes will have to take that into consideration.

### Definitional and Regulatory Discrepancies

These ambiguities have complicated the task of control over the arms trade. Compounding the problem is the lack of accord over what should be controlled. There is no internationally agreed-upon definition of what is an 'arms transfer', and so export regulations vary widely from state to state. Some define 'arms' narrowly to include lethal equipment only; others incorporate support equipment; fewer embrace defence-production facilities, technical assistance, or training; and still fewer consider dual-use components in this category.[73] Disagreement arises even over the definition of weapon types. The Chinese, for instance, consider their M-11 missile, with a reported range of about 180 miles, as a short-range missile, although the United States considers it a medium-range missile covered by the MTCR agreement.[74] These definitional asymmetries have been a source of some acrimony, fuelling the still-unresolved dispute as to what should and should not be regulated.

### Ambiguous and Conflicting Arms-Control Goals

Perhaps most intractable, however, are the ambiguous and con-flicting goals of the arms-control proposals themselves, past and present. None directly addresses the questions: 'Who are arms controls designed to protect?' 'Arms control for whose benefit?' The classic arms-control goals—to reduce the probability of war or its destructiveness should war occur—are too vague to be useful policy guidelines and are open to wide interpretation. To be effective, policy-makers must be clear about their priorities and

---

[72] U. Rubin, 'How Much Does Missile Proliferation Matter?', *ORBIS* (winter 1991), 29–39, at pp. 34–5.
[73] For a country-by-country analysis of arms-export regulations, see Anthony (ed.), *Arms Export Regulations*.
[74] 'China Said to Sell Parts for Missile', *New York Times*, 31 Jan. 1992.

focus the control-effort accordingly. They must decide whether the intent of arms-control accords is to safeguard Third World countries from other aggressive Third World countries? To shield the industrialized world from Third World threats? Or to protect industrialized countries from each other? Clearly goals determine the type of arms-control measures to be adopted. Weapons that are destabilizing in one regional context may not be in another. It is relatively easy to give lip-service to the general concept of arms control, and apparently many states cynically do so; but, unless the real purpose of arms control is articulated and agreed upon, it cannot be translated into implementable policy.

### A BALANCE SHEET: PROSPECTS FOR THE FUTURE

In view of these constraints, expectations for future arms-control measures must remain modest. As a practical matter, arms-control regimes do not augur the complete cessation of arms transfers. As in other foreign-policy areas, an equilibrium will have to be sought between the self-defence requirements of recipients, the economic exigencies of the suppliers, and the desire of the public and politicians to restrict transfers that threaten to destabilize regional balances. The best to be hoped for are regional or global agreements to limit the export of specific types of weapons on a case-by-case basis and general accords to control the flow of arms to belligerents.

Ultimately, however, it will be the interests, influence, and resolve of the United States that will determine the character and effectiveness of international arms-control measures. The deciding factor will be its political will. Above all, if the United States is to play the role of initiator and enforcer, even modest success will require a perception on the part of the leadership that arms-transfer limitations are in the long-term national interest of the United States. It will also require a significant expenditure of US political and economic collateral, as its recent attempts to contain nuclear proliferation demonstrate.[75]

Intelligence capabilities will also be critical here, to verify

---

[75] 'China Undercuts US Anti-Atom Effort on Korea', *New York Times*, 15 Nov. 1991; 'US Nuclear Technology Tactics Vex Iran', *New York Times*, 18 Nov. 1991; 'Algerian Reactor Came from China', *New York Times*, 15 Nov. 1991.

supplier and recipient compliance. Monitoring the shipment of equipment through technical means—satellite photographs, message intercepts, etc.—is relatively straightforward and unambiguous. Tracking the development and production of weapons in conflict regions will be more difficult. The failure to gauge the magnitude of Iraq's nuclear, chemical, and conventional programme, however, underscores this need.

More demanding still will be the requirement for broader, creative analysis to assess not only what is being produced and transferred, but how it will be used and what its strategic significance is over the long term. Grasping interrelationships among the diverse items being transferred from different sources to create a coherent picture of intention will not be easy, and will require a major commitment of intelligence-gathering and analytical resources.

Furthermore, if the United States, as the major producer of advanced technologies, hopes to contain the proliferation of advanced weaponry, it will have to keep better account of its own exported components and the use to which they are put by other countries. Third-party transfer restrictions can work only if the United States has good knowledge of which significant components have been exported, the foreign weapon systems in which they have been integrated, and whether the buyer is exporting those weapons to third parties. This, too, is a task that requires good intelligence. In an age of increased international co-operation, joint ventures, mergers, buy-ins, etc., it will become increasingly difficult to define what legally constitutes a US military product of, for that matter, a US defence company. But more needs to be done in this area.

Equally important and difficult to accomplish will be reaching consensus among suppliers and recipients on what to do when parties violate agreed-upon arms-limitation principles. Arms-control efforts in the past have been frustrated by lack of effective enforcement. Unless some penalties for violators are instituted and implemented, compliance for even minor conventional arms-transfer restraints will be difficult to achieve. Similarly, compensation measures to mitigate the costs of acquiescence to arms control may have to be established in some cases.

In spite of these obstacles, however, if we do not expect too much, there is cause for guarded optimism. Given the military technological revolution, a world economic recession, and the changing political order, an international climate of opinion is developing which may foster supplier–recipient arms restraint, and encourage states to perceive some form of arms control as in their own national interest. We have arrived at a unique moment in history, when the configuration of political and economic forces makes possible reduced military transfers.

The United States is positioned to make a difference. The intractable problem, of course, is that arms controls, like arms transfers, are foreign-policy instruments which are subject to conflicting political forces. For the United States, its perceived security interests may, at times, conflict with its commitment to restrain arms transfers—producing an element of policy schizophrenia that will confound friend and foe alike. It will take time and patience on everyone's part, but the balance appears to be shifting in favour of further regulation of the conventional arms trade.

# Conclusions

## EFRAIM KARSH, MARTIN NAVIAS, AND PHILIP SABIN

ON 14 April 1992 Iraq's main nuclear-weapons production complex at Al-Atheer, some forty kilometres south of Baghdad, was demolished by the IAEA inspectors. It was at this plant that Saddam Hussein had hoped to develop his nuclear device, and its destruction represented the culmination of an unprecedented international effort to neutralize Iraq's non-conventional power.

While it is generally accepted that Baghdad's nuclear-development project has suffered a devastating setback following the 1991 Gulf War, it is less clear whether the international non-proliferation effort has succeeded in completely and permanently removing Iraq from the list of developing states intent upon acquiring nuclear weapons. The Iraqis have always argued that their nuclear programme was designed for purely peaceful purposes, and only after the war, following revelations by deserting Iraqi nuclear scientists and several IAEA inspection tours to Iraq, did it transpire that Iraq's nuclear programme had been far more advanced and extensive than previously assumed, and that it had survived the allied attacks largely intact.

Moreover, to Saddam Hussein nuclear weapons have always meant much more than the 'great equalizer'. They have been a personal obsession—a symbol of Iraq's technological prowess, a prerequisite for regional hegemony, the ultimate guarantee of absolute security. Hence, salvaging whatever he can from his deadly arsenal is, in his eyes, a matter of life and death. It required eleven visits by IAEA inspection teams to Baghdad before Saddam relented and agreed to the destruction of Al-Atheer, together with the neighbouring Al-Hateen high-explosive complex, and his scientists have been less than forthcoming with regard to all procurement details of components, materials, and design know-how. This could form the basis of a later programme or be sold or transferred to other countries in the region or elsewhere. No wonder that, in June 1992, the Deputy Head of the IAEA, Maurizio Zifferero, expressed the fear that, while 'for the time being our inspectors have cut the head off efforts to turn Iraq into a nuclear threat... [the Iraqis] have got the know-how and the people, so it is only a matter of time before they could try to make a bomb'.[1]

While many in the West may dismiss such fears as overrated, Middle Easterners are unlikely to take a similarly relaxed view. For them, the possibility that Baghdad is still holding back on technology is disturbing, not only because it contains the seeds of a revived Iraqi programme, but also because it casts heavy doubts about the efficacy of future verification and arms-control regimes. If anything, the 1991 Gulf War has underscored how little the international community knows about the scope and capacity of the highly secretive Middle Eastern non-conventional-weapons programmes.

The unprecedented international co-operation in dismantling Iraq's non-conventional weapons was made possible by a unique convergence of regional and international conditions, which may not easily recur in the future. These ranged from the brutal nature of the Iraqi regime, manifested in domestic repression and external aggression, through the strategic and economic importance of the Middle East and the Persian Gulf, to the momentous events in Eastern Europe and the Soviet Union and the consequent diminution in great-power rivalry in the Third World. And yet, even this exceptional measure of global co-operation, forged during the Gulf conflict of 1990–1, and maintained—with great difficulty and diminishing efficiency—in the wake of the conflict, has encountered formidable obstacles and its result has been less than complete.

Hence, despite continuing efforts to expand and enhance verification procedures, it is questionable whether any arms-control regime can provide a full guarantee that all programmes and weapons are accounted for. This, in turn, may drive regional actors to continue their long-standing reliance on deterrence, rather than on multilateral arrangements. At the conventional

[1] *The Times*, 17 June 1992.

level such development may be tolerable, as conventional weapons do not threaten the existence of the entire system as such. However, in the non-conventional sphere, where the stakes involved are far higher, the demands for knowledge must be more stringent.

It has been argued in this book that traditional assessments of cost–benefit and military efficacy are of only limited utility in informing upon regional decisions to acquire non-conventional weapons. Furthermore, the 1991 Gulf War has done little to undermine the incentives of regional states to seek such weaponry. To the contrary, the general fear of Iraq's resort to non-conventional warfare underlined the symbolic and potentially practical significance of these weapons. The Iraqi missile strikes against Israel, and Washington's less than friendly attitude towards the Jewish State in the wake of the war, reinforced Israeli perceptions of a continuing hostile environment and fears that, in a future conflict, Israel might find itself more isolated than ever before. At the same time, the sensitivity of the Israeli population to conventional missile strikes demonstrated to its regional protagonists the potential efficacy of missiles, in particular when coupled with non-conventional warheads. Iran, especially, which is seeking to fill the power vacuum left by the neutralization of Iraq, seems to have taken note of these lessons and is investing heavily in missile and non-conventional-weapons projects.

It is not simply the strong incentives to acquire non-conventional weaponry that help underline the difficulties in ensuring agreement about arms control, but the asymmetry in motivation that underpins proliferation. In Israeli strategic thinking nuclear weapons are means of last resort, the so-called 'Samson's Option', designed to offset Arab massive conventional superiority in a situation where Israel's quality can no longer match enemy quantity. In addition, they are vital elements in deterring enemy non-conventional weaponry—a task that, in the Israeli view, cannot be entrusted to inevitably imperfect arms-control regimes. Conversely, Arab and Iranian non-conventional-weapons programmes are in great part responses to the Israeli arsenal, whose role they do not believe to be purely deterrent or 'last resort'. They are also a means of compensating for the failure to match Israel at the conventional level, and, not least, an instrument to be used in intra-Arab and Iranian–Arab hostilities.

As Israel approaches the historical juncture of significant

territorial concessions in return for peace, it is unlikely to willingly surrender its ultimate weapons, particularly since they may be viewed as a trade-off for the loss of strategic depth. In such circumstances, the Arab countries and Iran will most probably retain their chemical-weaponry or nuclear-weapons development efforts. But, even if Israel did the unthinkable and unilaterally surrendered its non-conventional weaponry, it is doubtful whether others in the region would follow suit. Quite the reverse, in fact. It is arguable that such a move would only enhance the desirability of non-conventional weapons amongst other regional states, as there would be no comparable system to deter employment.

The different rationales of non-conventional weaponry and programmes in the strategies of the various protagonists, the deeply entrenched distrust between them, and the lack of confidence in verification procedures means that, despite the large amount of effort devoted to arms control in the Middle East, it is unlikely that these weapons will be negotiated away, at least not in the short and medium terms. The Middle East may probably be one of the last regions where these types of international regimes will be implemented.

This scepticism notwithstanding, there is a general consensus among the contributors to this volume that efforts to constrain non-conventional regional proliferation will have to be made. Whatever the prospects for success, the problem cannot be ignored. The destructiveness of the weapons being proliferated, the volatility of the region, and the fear that future conflicts may not necessarily be confined to the Middle East alone, all mean that extra-regional powers cannot disengage themselves from what is happening in the area.

The key point here is to get the priorities right and, first and foremost, to recognize the pre-eminence of political solutions over arms-control arrangements. As has been demonstrated in the context of East–West relations, significant reductions in armaments followed superpower *détente*, not the other way around. Where hostility is intense and distrust is great, the prospects for massive reductions in weaponry remain slim. In the case of the Arab–Israeli conflict—or, for that matter, the Iraqi–Iranian rivalry—the political aspects of the disputes need to be seriously addressed before significant efforts are invested in getting the parties to reduce their arsenals. The greater the diminution in mutual threat perceptions, the stronger the readiness to control the proliferation of non-conventional, and conventional, weaponry.

Confidence-building measures, if carefully constructed, can serve to buttress such readiness. Through such means as delineation of red lines, limitations on deployment and weapons-testing, and the initiation of multilateral discussions on non-conventional arms control, they can help prevent war by miscalculation. Since discussions between Arabs and Israelis are already under way, these subjects should already be being drawn into the talks.

As the negotiations process advances, effort should be made to establish demilitarized zones, peace-keeping forces, enhanced surveillance and verification measures, as well as an Israeli–Palestinian security regime. Once a settlement to the Arab–Israeli dispute has been reached, more ambitious objectives should be sought. These would include agreements on nuclear, chemical, and biological weapons, conventional-force reductions, inspection of arms-productions sites, and the setting-up of regional arms-supply agreements.

Nevertheless, as pointed out in this volume, it would be extremely naïve to expect the resolution of the Arab–Israeli conflict to remove all sources of regional instability. For all its severity, this conflict is merely one hotbed in the tangled and interdependent web of Middle Eastern volatility. Religious fundamentalism, ethnic and national irredentism, socio-economic schisms, territorial disputes, and megalomaniacal tendencies, as well as personal animosity between state leaders, are likely to bedevil the region for some time to come.

How these problems are going to be addressed, and what consequences they are likely to entail for the proliferation of non-conventional weapons, remains to be seen. It is clear, however, that the main task facing extra-regional powers is to promote political reconciliation wherever possible, while continuing to upgrade and expand their supplier controls. This objective is clearly within their reach, and, while it cannot by itself stop the regional arms race, it may slow it down and inject a measure of predictability into an unstable dynamic.

The importance of such measures cannot be overstated. While it is possible to live with some discrepancy between declaratory pronouncements about the need for arms control and actual conventional-arms-sales policies, this will not be possible in the area of non-conventional weapons, where the dangers are so much greater. If Middle Eastern states are to be convinced to relinquish their non-conventional arsenals, they will have to be fully confident that everybody is doing the same, and that nobody, in the region and beyond, is cheating.

# INDEX